NEW RESEARCH ON CHAOS AND COMPLEXITY

NEW RESEARCH ON CHAOS AND COMPLEXITY

FRANCO F. ORSUCCI
AND
NICOLETTA SALA
EDITORS

Nova Science Publishers, Inc.
New York

For permission to use material from this book please contact us:
Telephone 631-231-7269; Fax 631-231-8175
Web Site: http://www.novapublishers.com

Library of Congress Cataloging-in-Publication Data

New research on chaos and complexity / Franco F. Orsucci and Nicoletta Sala (editors).
p. ; cm.
Includes index.
ISBN 1-60021-275-1
1. Chaotic behavior in systems. I. Orsucci, Franco. II. Sala, Nicoletta.
[DNLM: 1. Nonlinear Dynamics. QA 427 N532 2006]
Q172.5.C45N49 2006
515'.39--dc22 2006017708

Published by Nova Science Publishers, Inc. ✢New York

CONTENTS

In: New Research on Chaos and Complexity ISBN 1-60021-275-1
Editors: F. Orsucci and N. Sala, pp. vii-xiii

PREFACE

Franco F. Orsucci
Clinical Health Psychology Sub-Department, University College,
London and Kent Institute of Medicine and Health Sciences,
University of Kent at Canterbury, UK

For his course is not round; nor can the Sunne
Perfit a Circle or maintaine his way
One inche direct; but where he rose to day
He comes no more, but with a cousening line,
Steales by that point, and so is Serpentine.

John Donne, An Anatomie of the World, 1611

A STATE OF THE ART

The ancient English of these verses brings a remarkable insight we ought to the poet John Donne. This poem highlights how the consciousness of complexity has been present for very long times in human cultures, even long time before these verses. It is quite recently, however, that it has become suitable of a scientific approach.

John von Neumann, circa 1950, affirmed:

"All stable processes, we shall predict. All unstable processes, we shall control." (cited in Dubè, 2000).

But, in his 1985 Giord Lectures, Freeman Dyson (1988) expressed his quite different opinion:

> "A chaotic motion is generally neither predictable nor controllable. It is unpredictable because a small disturbance will produce exponentially growing perturbation of the motion. It is uncontrollable because small disturbances lead only to other chaotic motions and not to any stable and predictable alternative."

Was Von Neumann's a mistake to imagine that every unstable motion could be nudged into a stable motion by ***small pushes and pulls applied at the right places*****?** If chaos is one of the possible marriages between order and disorder, habit and life, how much is it possible ***taming complex systems***?

The enterprise is still at the beginning, and the science of complexity, at this stage of development, still resembles a sea of ignorance with some small islands where results are known and applicable. The important thing is that now we accept that this sea exists and we can explore it.

This is well represented in the sketch below that we owe to Francisco Varela (1991) and Thomas Schreiber (Schreiber, 1999).

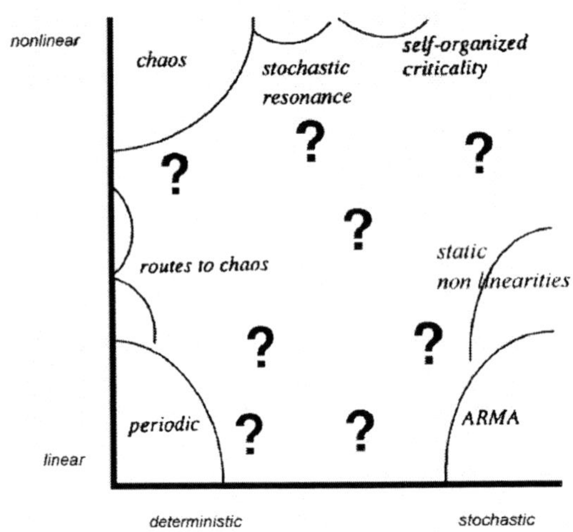

A state of the art in complexity theory (Varela, 1991; Schreiber, 1999)

Complexity in Metaphor

These small islands of knowledge called ***chaos, SOC or stochastic resonance*** are like candles in the darkness: we can finally have intuitions of the elephant's whole shape.

There is a Sufi tale called ***The Elephant in the Dark***:

> Some Hindus had brought an elephant for exhibition and placed it in a dark house. Crowds of people were going into that dark place to see the unknown beast. Finding that ocular inspection was impossible, each visitor felt it with his palm in the darkness.
>
> The palm of one fell on the trunk.
> 'This creature is like a water-spout,' he said.
> The hand of another lighted on the elephant's ear.
> To him the beast was evidently like a fan.
> Another rubbed against its leg.
> 'I found the elephant's shape is like a pillar', he said.
> Another laid his hand on its back.
> 'Certainly this elephant was like a throne', he said.

16th century Chinese painting about the story.

"The sensual eye is just like the palm of the hand. The palm has not the means of covering the whole of the best. The eye of the Sea is one thing and the foam another. Let the foam go, and gaze with the eye of the Sea. Day and night foam-flecks are flung from the sea: amazing! You behold the foam but not the Sea.

We are like boats dashing together; our eyes are darkened, yet we are in clear water" (Rumi, 1995).

The enterprise is crucial but not new. Classical knowledge was called ***Philosophia Naturalis***, ***Naturwissenschaften*** or ***Natural Philosophy***, until it had been fragmented and oversimplified in many sub-disciplines. Natural Philosophy, instead, had been rooted on the integration of different ways to approach a reality recognized as complex and multi-ordered.

The Sufi tale is clear: when the object is so large and complex, your perspectives can be partial and misleading. Sometimes the attempts to handle complexity have produced just the abuse of Occam's tools, with some risks for the epistemological survival of the object.

Dynamical Systems Theory

The gradual emergence of a set of formal and methodological tools called Dynamical Systems Theory, or Complexity Theory could finally make the scene different. This discipline has been also called Nonlinear Science as a marker of the shift in scientific paradigms (Kuhn, 1996). The name by exclusion might seem surprising for someone, as Stanislaw Ulam said: "Calling a science 'nonlinear' is like calling zoology 'the study of non-human animals' ". Anyway, the shift in scientific paradigms has been strong: a real scientific revolution.

Almost every system is complex, dynamical and nonlinear by nature. It is the limit of our approaches or our deliberate choice that makes us see them as linear. Yet the distinction has been necessary in the natural sciences, which became so accustomed to linear systems, because they were more treatable: linearization could "tame" their "wild" complexities.

The discovery of the possible scientific study of dynamical behaviors unrestricted by linearity is one of the greatest scientific revolutions of all times. It is becoming even a revolution in our everyday perception of reality, as trees and lightning become scientific objects under the name of fractals, just as cubes and cones have been for centuries.

It is clear that Maxwell and Boltzmann, the founders of statistical physics, were acutely aware of the property of sensitivity to initial conditions and its consequences. Not before Poincaré (1892) however, could ascertain the existence of this property in a system with few degrees of freedom, namely the reduced 3-body problem. In the continuing history of nonlinear dynamical systems, the first evidence of physical chaos is associated with the name of Edward Lorenz (1963; 1994) whose discovery of the first strange attractor in a simplified meteorological model containing only 3 state variables has led to a remarkable explosion in the study of chaos and its properties.

More recent years have seen the definition of a new frontier in complexity studies: the theory and application of control and synchronization. This Journal is proud to include in its Board many of the founders of this new wave of nonlinear studies.

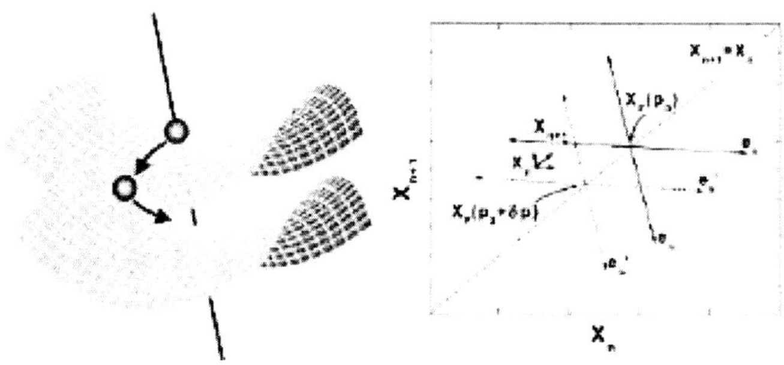

Local geometry of control: left 2D saddle dynamics and right linearization of the stable and unstable manifolds (Dubè, 2000)

VARIATION AND SELECTION

Chaos theory becomes also a crucial way to understand some deep implications of Darwin's research on biological laws. If chaos is a source of optimal variation, targeting desirable states within chaotic attractors is a preliminary phase of selection and co-evolution. One of the major problems in the above process is that one can switch on the control only when the system is sufficiently close to the desired behavior. This is warranted by the ergodicity of chaos regardless of the initial condition chosen for the chaotic evolution, but it may happen that the small neighborhood of a given attractor point (target) may be visited only infrequently, because of the locally small probability function.

This is just one of the many questions that are still open. David Ruelle (1994), almost ten years ago, summarized some methodological *caveats:* "Suppose that you have concocted a mathematical model in biology or economics; you put this model on your computer and you discover a Feigenbaum period-doubling cascade (...) is this result interesting?" He answered that probably it hasn't a lot of interest: you should care about the relation between your model and the real empirical situations. Real systems are not directly equivalent to computer models: "Computer study of a model is an important method of investigation, but the results can be only as good as the model".

However, the greatest challenge will remain for some time the application to complex biological systems: in particular to mind and brain dynamics (Freeman, 1999; Guastello, 1995; Orsucci, 2003).

The perspective of unifying the techniques of deterministic chaos control with a statistical description as a possible therapeutic strategy against dynamical diseases is the challenge for next years.

The following is a photo about the meeting between a man and some cetaceans, in a quasi topological and dynamical presentation of a coevolving interaction: a smart way to deal with big animals. It confirms that sometimes artists can find some metaphorical knowledge that scientists are trying to conquer in more formal ways (Verhulst, 1994).

Cetaceans and man play synchronized underwater (Colbert, 2002)

References

Colbert D (2002) *Ashes and Snow*, Venice: Biennale Monographs and Catalogues.

Dubè JL; Desprès P (2000) The Control of Dynamical Systems - Recovering Order from Chaos, in Itikawa,Y (Ed.) *The Physics of Electronic and Atomic Collisions*, Woodbury, N.Y.: AIP Conference Proceedings.

Dyson F (1988), *Infinite in All Directions*, New York: Harper and Row Publishers.

Freeman WJ (1999). *How Brains Make up their Minds*, London: Weidenfeld & Nicolson.

Gardner H (1985) *The mind's new science, a history of the cognitive revolution.* New York: Basic Books.

Guastello SJ (1995) *Chaos, catastrophe, and human affairs: applications of nonlinear dynamics to work, organizations, and social evolution,* Mahwah NJ: Lawrence Erlbaum Associates.

Kuhn TS (1996) *The Structure of Scientific Revolutions*. Chicago, IL: University of Chicago Press.

Lorenz EN (1963) *J. of Atmos. Sci.* 20, 130 .

Lorenz EN (1994) *The Essence of Chaos (The Jessie and John Danz Lecture Series)*, University of Washington Press.

Orsucci F (2003) *Changing Mind: Transitions in Natural and Artificial Environments,* Singapore: World.Scientific.

Poincaré H (1892) *Les Methodes Nouvelles de la Mecanique Celeste*, Paris: Gauthier-Villars et fils 13, 1 .

Ruelle D, (1994) Where can one hope to profitably apply the ideas of chaos?, *Physics Today* 47, 7, 24-30.

Rumi JJ-D & Barks C (1995) *The Essential Rumi*. San Francisco, CA: Harper.

Schreiber T (1999) Interdisciplinary application of nonlinear time series methods, *Physics Reports* 308, 1-64.

Varela FJ, Thompson E & Rosch E (1991) *The Embodied Mind: Cognitive Science and Human Experience*, Cambridge, Mass: MIT Press.

Verhulst F (1994) Metaphors for psychoanalysis, *Nonlinear Science Today* 4 (1):1-6.

In: New Research on Chaos and Complexity ISBN 1-60021-275-1
Editors: F. Orsucci and N. Sala, pp. 1-3

Chapter 1

MEMORIAL: ILYA PRIGOGINE AND HIS LAST WORKS

Gonzalo Ordonez
Center for Studies in Statistical Mechanics and Complex Systems, The University of Texas at Austin, Austin, Texas 78712, USA

During his long and extremely fruitful career, Prof. Ilya Prigogine worked on many different subjects, from the theory of molecular solutions, to the theory of vehicular traffic and the big bang. He was one of the pioneers in the field of non equilibrium thermodynamics, especially with his work on dissipative structures, which showed that the increase of entropy, usually associated with increasing disorder, could also lead to self-organization and complexity in open systems.

He often spoke of unification between man and nature, connected through time in its creative role. He was inspired by Bergson, who said (I. Prigogine, Autobiographie, Florilège des Sciences en Belgique II, 1980):

> "The more deeply we study the nature of time, the better we understand that duration means invention, creation of forms, continuous elaboration of the absolutely new."

The study of time was a recurring theme in his scientific career. He kept working on this theme, and other problems in physics derived from it, throughout his last years. I had the great privilege of working with Prof. Prigogine during this period.

When talking about physics, he had as much enthusiasm as a freshly graduated student. He was always looking into the future, coming up with new problems to work on. This was quite consistent with his philosophical views on time. "The future is open," "we are only at the beginning" were common phrases he used.

One of the subjects that most interested Prof. Prigogine in his last years was the study of entropy and its connection to dynamics. Traditionally, this connection has been made through the introduction of supplementary assumptions or approximations. But many questions remain surrounding entropy: how to define entropy for dense systems, as well as for systems far from equilibrium? Prof. Prigogine believed that to answer these questions one should look more closely at the transition from the dynamical description, given by classical or quantum dynamics, to the thermodynamical description. This transition occurs at the limits of dynamics, when the solutions of equations of motion become so irregular that they cannot be written in a compact way. As described by Poincarè, they become "non-integrable." Prof. Prigogine and collaborators showed, however, that one can introduce new representations that yield thermodynamic behavior, without approximations. Quoting from his autobiography:

> "... I was prompted by a feeling of dissatisfaction, as the relation with thermodynamics was not established by our work in statistical mechanics, nor by any other method ... the question of the nature of dynamical systems to which thermodynamics applies was still without answer."

> "...If irreversibility does not result from supplementary approximations, it can only be formulated in a theory of transformations which expresses in "explicit" terms what the usual formulation of dynamics does "hide". In this perspective, the kinetic equation of Boltzmann corresponds to a formulation of dynamics in a new representation ... In conclusion: dynamics and thermodynamics become two complementary descriptions of nature, bound by a new theory of non-unitary transformations."

On this subject, together with collaborators he was able to make much progress during the last years. We could precisely define, at least in simple cases, a "microscopic entropy," derived from a non-unitary transformation. This entropy could measure the "age" of a system out of equilibrium.

Related to this, Prof. Prigogine thought he could find a new view on the "twin paradox."

This is a well know paradox in the theory of relativity, used to show that acceleration can lead to slower aging, due to relativistic time dilation. This fact prompted him to study the effects of acceleration on age, age defined through the microscopic entropy mentioned above. I am sure he had a deeper question in mind: what is the relation between the "geometric" time of relativity, and the "thermodynamic" time, connected with increasing entropy? In this context we considered first a non-relativistic situation, showing that acceleration can indeed lead to "rejuvenation." (here due to different causes than in the twin paradox). We started to write a paper on this. I prepared a draft, following many suggestions from Prof. Prigogine. The next step was for him to have a look on it, but unfortunately he became ill and passed away.

Here I want to say that Prof. Prigogine kept interest in his work until the very end. He even wanted to work on this paper while he was in the hospital. This was not uncharacteristic of him. In a previous, less serious occasion, he had to stay in the hospital for a few days. Students and colleagues would visit him to discuss physics. At some point we thought that we should bring a blackboard to his hospital room!

The editors of "Chaos and Complexity Letters" have kindly accepted to publish this paper on acceleration and entropy after Prof. Prigogine's death. As I mentioned, this paper really shows work in progress. I hope that it will draw attention to some of the subjects that Prof. Prigogine was working on, and which I think are worth pursuing further.

I believe Prof. Prigogine has made a deep, lasting contribution to our understanding of nature, giving us a glimpse on the mechanisms of self-organization, a key element in our understanding of life. For the people who knew him, he left as well unforgettable memories, marked by his great, warm human nature and contagious enthusiasm. As a tribute to him we can continue his work, guided by his aim of unification between different disciplines, unification between man and nature. As he said, time is not an illusion; the future is widely open for us to shape.

Gonzalo Ordonez
Austin, Texas, June 30, 2003.

In: New Research on Chaos and Complexity ISBN 1-60021-275-1
Editors: F. Orsucci and N. Sala, pp. 5-20

Chapter 2

ACCELERATION AND ENTROPY: A MACROSCOPIC ANALOGUE OF THE TWIN PARADOX

I. Prigogine and G. Ordonez
Center for Studies in Statistical Mechanics and Complex Systems,
The University of Texas at Austin, Austin, TX 78712 USA
andInternational Solvay Institutes for Physics and Chemistry,
CP231, 1050 Brussels, Belgium

Abstract

The twin-paradox described in relativity theory shows that acceleration leads to slower aging. Motivated by this, we consider the effects of acceleration on entropy. We consider a macroscopic, nonrelativistic analogue of the twin effect on a 2-D weakly coupled gas. We introduce a dynamical entropy ($\mathcal{H}$ function), which measures the "age" of the system. In previous papers we have considered the effect of rejuvenation by velocity inversion of every particle. Here we generalize our results by studying how the $\mathcal{H}$–function changes as a result of rotation of the velocities by a given angle. The rotation is the result of some acceleration. Therefore acceleration leads to entropy flow. The degree of "rejuvenation" of the system depends on the angle. As a special case, a rotation by $\pi/2$ approximately resets the $\mathcal{H}$-function to its initial value. In thermodynamic terms, there is a compensation between entropy production and entropy flow.

1 Introduction

The twin paradox is explained by acceleration on the moving twin. It has been verified by the time delay of unstable particles, related to relativistic field theory. Still, it is not clear why this would have an effect on chemical reactions or living material. To study this one would require a relativistic formulation of nonequilibrium physics. Therefore it is not without interest to consider a non-relativisitic effect of acceleration on aging. It

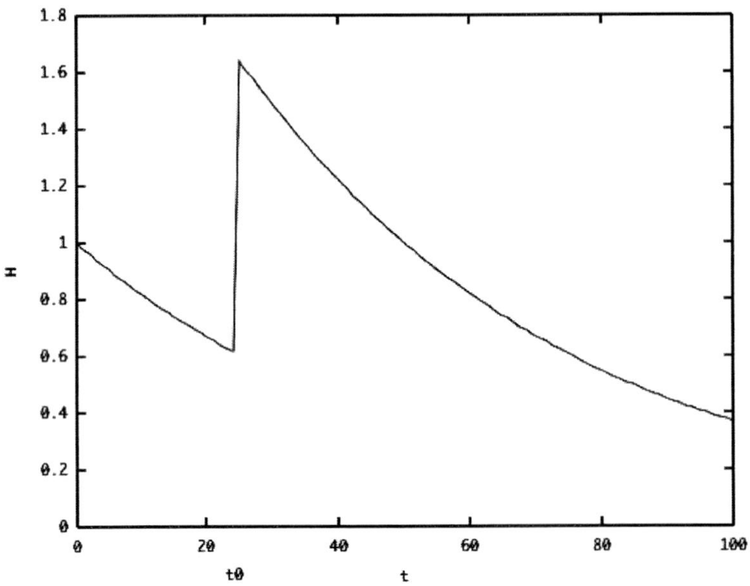

Figure 1: Schematic plot of the Lyapounov function $\mathcal{H}(t)$, showing the effects of velocity inversion at a given time t_0. Velocity inversion creates correlations and causes $\mathcal{H}$ to jump up.

is generally accepted that the age of such systems is related to entropy. We define non-equilibrium entropy through an $\mathcal{H}$-function, analogous to Boltzmann's $\mathcal{H}$-function, but incorporating correlations [2]. When we start from a non-equilibrium state, the time evolution leads to approach to equilibrium through relaxation processes. If we then invert velocities we have a highly complex behavior corresponding to a time-reversed evolution. This corresponds to an injection of "negative entropy," leading to a rejuvenation of the system. Post-collisional correlations are turned into pre-collisional correlations, and this leads to a "jump" or discontinuity in the $\mathcal{H}$-function. After the jump the $\mathcal{H}$-function continues to decrease, due to the decay of the pre-collisional correlations (see Fig. 1). This is the answer we give to the Lochsmidt paradox [2]. In thermodynamics the change of entropy is

$$dS = d_i S + d_e S \tag{1}$$

where $d_i S$ is the entropy production and $d_e S$ is the entropy flow. The inversion of velocities corresponds to a flow of entropy. In a 2-D system velocity inversion corresponds to a rotation of the velocity of every particle by an angle $\phi = \pi$. The question we want to study is the generalization of this problem to arbitrary angles ϕ for velocity rotations. We will study a 2-D, weakly interacting classical gas. We will estimate what is the effect of finite velocity rotations, by an angle ϕ, on the $\mathcal{H}$-function. As we will see rejuvenation will now depend on the angle. So, already in a non-relativistic setting, acceleration has an effect on age. This is of course a highly idealized situation because we need a force which would turn all the velocities by the same angle ϕ. But we could consider systems for which we can turn the velocities in a finite region. Then the effect would disappear after a time depending on the extension of the region.

2 Weakly coupled gas

We consider a classical 2-D gas with Hamiltonian

$$H = \sum_{n=1}^{N} \frac{\mathbf{p}_n^2}{2m} + \sum_{n<j}^{N} \lambda V(|\mathbf{r}_n - \mathbf{r}_j|) \tag{2}$$

We write the potential energy as

$$V(r) = \left(\frac{2\pi}{L}\right)^2 \sum_{\mathbf{k}} V_k \exp(i\mathbf{k}\cdot\mathbf{r}) \tag{3}$$

where $k = |\mathbf{k}|$ and L is the size of the system. We are interested in the thermodynamic limit $N \to \infty$, $L \to \infty$, keeping a finite concentration

$$c = N/L^2 > 0 \tag{4}$$

To simplify our discussions, we will consider dilute gas with weak coupling

$$\lambda \ll 1 \tag{5}$$

Hereafter we will use units where $m = 1$ and we will work with the velocities $\mathbf{v}_n = \mathbf{p}_n/m = \mathbf{p}_n$ rather than with the momenta. Ensembles

$\rho(\mathbf{r}_1, \ldots \mathbf{r}_N, \mathbf{v}_1, \ldots \mathbf{v}_N, t)$ or $\rho(t)$ in short, evolve according to the Liouville equation,

$$i\frac{\partial}{\partial t}\rho(t) = L_H\rho(t), \qquad \rho(t) = \exp(-iL_Ht)\rho(0) \tag{6}$$

where

$$L_H = -i\sum_{n=1}^{N}\left(\frac{\partial H}{\partial \mathbf{v}_n}\cdot\frac{\partial}{\partial \mathbf{r}_n} - \frac{\partial H}{\partial \mathbf{r}_n}\cdot\frac{\partial}{\partial \mathbf{v}_n}\right) \tag{7}$$

with

$$\begin{aligned}\frac{\partial}{\partial \mathbf{v}} &= \left(\frac{\partial}{\partial v_x}, \frac{\partial}{\partial v_y}\right)\\ \frac{\partial}{\partial \mathbf{r}} &= \left(\frac{\partial}{\partial r_x}, \frac{\partial}{\partial r_y}\right)\end{aligned} \tag{8}$$

We decompose the Liouvillian as $L_H = L_0 + \lambda L_V$ where

$$L_0 = -i\sum_{n}^{N}\mathbf{v}_n\cdot\frac{\partial}{\partial \mathbf{r}_n} \tag{9}$$

$$L_V = \sum_{n<j}^{N}\sum_{\mathbf{k}} V_k e^{i\mathbf{k}\cdot(\mathbf{r}_n - \mathbf{r}_j)} i\mathbf{k}\cdot\left(\frac{\partial}{\partial \mathbf{v}_n} - \frac{\partial}{\partial \mathbf{v}_j}\right) \tag{10}$$

We consider initial ensembles that depend only on the velocities

$$\rho(0) = P^{(0)}\rho(0) \tag{11}$$

where $P^{(0)}$ is the projector into homogenous distributions:

$$P^{(0)}\rho = \frac{1}{L^{3N}}\int d^2r_1 \ldots d^2r_N\, \rho \tag{12}$$

We denote the complement projector as

$$Q^{(0)} = 1 - P^{(0)} \tag{13}$$

The unperturbed Liouville operator satisfies the property

$$L_0P^{(0)} = P^{(0)}L_0 = 0 \tag{14}$$

because functions in the $P^{(0)}$ subspace are independent of the positions of the particles.

3 The Λ transformation

In the following sections we will introduce an "entropy" operator to define the "age" of our system. This operator is constructed starting with a transformation Λ we have introduced previously. In this section we will give an overview of the Λ transformation. More details can be found in Refs. [2, 4]. We define inner products between functions of phase-space variables and ensembles as

$$\langle\langle f|\rho\rangle\rangle = \int d^2r_1 \dots d^2v_N f^*(\mathbf{r}_1 \dots \mathbf{v}_N)\rho(\mathbf{r}_1 \dots \mathbf{v}_N) \tag{15}$$

With this inner product we define Hermitian conjugation as usual

$$\langle\langle f|O|\rho\rangle\rangle = \langle\langle \rho|O^\dagger|f\rangle\rangle^* \tag{16}$$

Canonical transformations U are unitary:

$$U^\dagger = U^{-1} \tag{17}$$

For integrable systems in the sense of Poincaré, we can construct by perturbation series or otherwise a canonical transformation U that eliminates the interactions. We define new phase space variables

$$\begin{aligned} \tilde{\mathbf{r}}_n &= U^\dagger \mathbf{r}_n \\ \tilde{\mathbf{v}}_n &= U^\dagger \mathbf{v}_n \end{aligned} \tag{18}$$

such that the Hamiltonian is only a function of the new velocities $H(\mathbf{r}_1 \dots \mathbf{v}_N) = \tilde{H}(\tilde{\mathbf{v}}_1 \dots \tilde{\mathbf{v}}_N)$. In terms of the new variables, the equations of motion are enormously simplified. For integrable systems the operator $\tilde{L}_0 = UL_HU^{-1}$ gives the Liouville operator of free particles, so we have (see Eq. (14))

$$P^{(0)}\tilde{L}_0 = \tilde{L}_0 P^{(0)} = 0 \tag{19}$$

One can construct U using standard perturbation theory. For example, acting on the homogenous subspace $P^{(0)}$ we have

$$P^{(0)}U = P^{(0)} + P^{(0)}\lambda L_V \frac{1}{-L_0} Q^{(0)} + O(\lambda^2) \tag{20}$$

$$U^{-1}P^{(0)} = P^{(0)} + Q^{(0)} \frac{1}{-L_0} \lambda L_V P^{(0)} + O(\lambda^2) \tag{21}$$

However, the situations where this can be done are exceptional. Most systems are non-integrable, due to divergences (vanishing denominators) in the perturbation expansion of U, caused by resonances. Denominators appear as

$$\frac{1}{\omega} \tag{22}$$

where, e.g., $\omega = \mathbf{k}\cdot\mathbf{v}$. We have shown in Refs. [2, 4] that one can remove the divergences through regularization of denominators

$$\frac{1}{\omega} \Rightarrow \frac{1}{\omega \pm i\epsilon} = \mathcal{P}\frac{1}{\omega} \mp \pi i\delta(\omega) \tag{23}$$

The denominators are interpreted as distributions (generalized functions). This leads to a non-unitary transformation Λ, which replaces U. We no more obtain a description in terms of free particles. Instead, we obtain a "kinetic" description with broken time-symmetry. The sign of $\pm i\epsilon$ is chosen depending on the types of transitions (from lower to higher correlations or vice versa), which corresponds to a "dynamics of correlations" [1, 8, 4]. In terms of the new variables, we obtain now probabilistic, irreversible equations. It is remarkable that Λ is invertible. We can go back and forth between the dynamic description and the kinetic description. Instead of the operator $\tilde{L}_0$ we introduce the operator

$$\tilde{\theta} = \Lambda L_H \Lambda^{-1} \tag{24}$$

The Λ transformation satisfies the block-diagonal property

$$P^{(0)}\tilde{\theta} = \tilde{\theta}P^{(0)} \equiv \tilde{\theta}^{(0)} \tag{25}$$

which replaces Eq. (19). $\tilde{\theta}$ is now a collision operator, as used in kinetic theory. In the perturbation expansion we have now

$$P^{(0)}\Lambda = P^{(0)} + P^{(0)}\lambda L_V \frac{1}{i\epsilon - L_0} Q^{(0)} + O(\lambda^2) \tag{26}$$

$$\Lambda^{-1}P^{(0)} = P^{(0)} + Q^{(0)}\frac{1}{i\epsilon - L_0}\lambda L_V P^{(0)} + O(\lambda^2) \tag{27}$$

which are the extensions of Eqs. (20), (21). Note that the sign of ϵ is the same in both expressions. Due to this, Λ is no more unitary:

$$\Lambda^{-1} \neq \Lambda^{\dagger} \tag{28}$$

Instead, it is "star-unitary" [2, 4]

$$\Lambda^{-1} = \Lambda^{\star} \tag{29}$$

One of the most interesting features of the Λ transformation is that it allows us to introduce an "entropy" operator or, more precisely speaking, an $\mathcal{H}$-function.

4 $\mathcal{H}$-function

In our earlier work [2, 7] we have shown that if there exists a dynamical entropy it must be an operator. For systems that present Poincaré resonances, one can introduce the entropy operator

$$M = \Lambda^{\dagger}\Lambda \tag{30}$$

As shown in Ref. [2], the average of M is a positive, monotonic function of time. In other words, it is a Lyapiunov function. In this sense we can associate M with a generalized entropy. If Λ were replaced by the unitary transformation U, we would find that $M = U^{\dagger}U = 1$. So for integrable systems our entropy remains constant, just like Gibbs' entropy. The operator M depends on all the particles of the system. One can introduce reduced operators depending only on a limited number of particles [8]. We will consider the reduced operator

$$M_1 = \Lambda^{\dagger}|f_1\rangle\!\rangle\langle\!\langle f_1|\Lambda \tag{31}$$

where

$$\langle\!\langle f_1|\rho\rangle\!\rangle = \int d^2r_1 \ldots d^2r_n \int d^2v_1 \ldots d^2v_n\, f_1(\mathbf{v}_1)^* \rho(\mathbf{r}_1, \cdots \mathbf{r}_N, \mathbf{v}_1, \cdots \mathbf{v}_N) \tag{32}$$

is the reduced, one-particle velocity distribution function. We define the $\mathcal{H}$-function as

$$\mathcal{H}(t) = \langle\!\langle \rho(t)|M_1|\rho(t)\rangle\!\rangle \tag{33}$$

As we will show below, this is a Lyapounov function of the system [8].[1] For simplicity we will consider functions f_1 that depend only on the magnitude of the velocity

$$f_1(\mathbf{v}) = f_1(v) \tag{34}$$

[1] A very simple example of our $\mathcal{H}$-function has been given in Refs. [6, 3] for the quantum Friedrichs model, which consists of a discrete state ("atom") coupled to a continuum of field modes ("photons").

where $v = |\mathbf{v}|$. We write the $\mathcal{H}$-function as

$$\mathcal{H}(t) = \mathcal{A}^2(t) \tag{35}$$

where

$$\mathcal{A}(t) = \langle\langle f_1|\Lambda|\rho(t)\rangle\rangle \tag{36}$$

The $\mathcal{H}$-function is a Lyapounov function because it is positive and it is non-increasing for all t. Indeed we have (see Eq. (24))

$$\mathcal{A}(t) = \langle\langle f_1|\Lambda e^{-iL_H t}|\rho(0)\rangle\rangle = \langle\langle f_1|e^{-i\tilde{\theta}t}|\tilde{\rho}(0)\rangle\rangle \tag{37}$$

where

$$|\tilde{\rho}(0)\rangle\rangle = \Lambda|\rho(0)\rangle\rangle \tag{38}$$

Noting that $\langle\langle f_1| = \langle\langle f_1|P^{(0)}$ we obtain (see Eq. (25))

$$\mathcal{A}(t) = \langle\langle f_1|e^{-i\tilde{\theta}^{(0)}t}|\tilde{\rho}(0)\rangle\rangle \tag{39}$$

As shown in Refs. [2, 9] the operator $\tilde{\theta}^{(0)}$ is purely imaginary,

$$[\tilde{\theta}^{(0)}]^\dagger = -\tilde{\theta}^{(0)} \tag{40}$$

and as shown below it has the property

$$i\tilde{\theta}^{(0)} \geq 0 \tag{41}$$

which breaks time-symmetry in Eq. (39). This shows that $\mathcal{A}(t)$ is non-increasing. This function may be interpreted as a renormalized velocity distribution, corresponding to the velocity distribution of dressed quasiparticles. The evolution of quasiparticles is strictly Markovian and irreversible. In contrast, the time evolution of the original particles (bare particles) is not Markvian, because of the existence of dressing processes. The Λ transformation thus allows us to separate dressing processes from irreversible processes. Age corresponds to the evolution of dressed particles. To show Eq. (41), we note that due to the similitude relation in Eq. (24), the operators $\tilde{\theta}$ and L_H share the same eigenvalues [8]. The eigenvalues of L_H are given by the singularities of the resolvent operator

$$R(z) \equiv \frac{1}{z - L_H} \tag{42}$$

It is well-known [1] that, depending on the analytic continuation of the resolvent (from the upper to the lower half-plane of z or viceversa) all the complex singularities are either on the lower or on the upper half plane. We choose the analytic continuation from the upper to the lower half-planes, since we are interested in extracting contributions to the time evolution operator $\exp(-iL_H t)$ that decay for $t > 0$. In this branch of the resolvent, all the eigenvalues of L_H (and hence of $\tilde{\theta}$) are thus either real or on the lower half-plane. This proves Eq. (41). Our $\mathcal{H}$-function extracts the exponential decay processes during the approach to equilibrium of the system for $t > 0$. As discussed in [2], if we perform a velocity inversion (equivalent to a time inversion) we have the change $\tilde{\theta} \Rightarrow -\tilde{\theta}$ in the exponential in Eq. (39). Indeed, introducing the velocity inversion operator I we have

$$IL_H = -IL_H \tag{43}$$

After the inversion the $\mathcal{A}$-function changes as

$$\mathcal{A}(t) \Rightarrow \mathcal{A}_I(t) = \langle\langle f_1|\Lambda I|\rho(t)\rangle\rangle \tag{44}$$

Due to the property (43), instead of exponential decay we have now exponential growth and $\mathcal{A}$ jumps (see Fig. 1). The jump in the $\mathcal{A}$-function after velocity inversion is related to the injection "negative entropy" that creates anomalous correlations between the particles [2]. We obtain a rejuvenation of the system. The more time we wait, the larger the negative entropy needed to perform the inversion. Indeed, $\mathcal{A}_I(t)$ grows exponentially with t. In a 2-D space, velocity inversions are equivalent to velocity rotations by an angle π. Next we will study the effects of velocity rotations for arbitrary angles ϕ.

5 Effects of velocity rotations

We introduce a velocity rotation operator R_ϕ acting on a velocity $\mathbf{v} = (v_x, v_y)$ as

$$R_\phi \mathbf{v} = (v_x \cos\phi + v_y \sin\phi, -v_x \sin\phi + v_y \cos\phi) \tag{45}$$

and on functions of the velocities

$$R_\phi F(\mathbf{v}_1, \mathbf{v}_2, \ldots \mathbf{v}_N) = F(R_\phi \mathbf{v}_1, R_\phi \mathbf{v}_2, \ldots R_\phi \mathbf{v}_N) \tag{46}$$

We will assume that the initial ensemble depends only on the magnitudes of all the velocities

$$\rho(\mathbf{r}_1, \ldots \mathbf{r}_N, \mathbf{v}_1, \ldots \mathbf{v}_N, 0) = \rho(\mathbf{r}_1, \ldots \mathbf{r}_N, |\mathbf{v}_1|, \ldots |\mathbf{v}_N|, 0) \tag{47}$$

This means that

$$R_\phi \rho(0) = \rho(0) \tag{48}$$

for any angle ϕ. After a velocity rotation the function $\mathcal{A}(t)$ in Eq. (36) changes as

$$\mathcal{A}(t) \Rightarrow \mathcal{A}_\phi(t) = \langle\langle f_1|\Lambda R_\phi|\rho(t)\rangle\rangle \tag{49}$$

or

$$\mathcal{A}_\phi(t) = \langle\langle f_1|\Lambda R_\phi e^{-iL_H t}|\rho(0)\rangle\rangle \tag{50}$$

Defining

$$L_H(\phi) \equiv R_\phi L_H R_\phi^{-1} \tag{51}$$

we have

$$\mathcal{A}_\phi(t) = \langle\langle f_1|\Lambda e^{-iL_H(\phi)t} R_\phi|\rho(0)\rangle\rangle \tag{52}$$

Using Eq. (48) we have then

$$\mathcal{A}_\phi(t) = \langle\langle f_1|\Lambda e^{-iL_H(\phi)t}|\rho(0)\rangle\rangle \tag{53}$$

To calculate the "rotated" Liouvillian $L_H(\phi)$ we use the definition (7). Noting that

$$R_\phi \frac{\partial}{\partial \mathbf{v}} f(\mathbf{v}) = \frac{\partial}{\partial \mathbf{v}_\phi} f(\mathbf{v}_\phi) = \frac{\partial}{\partial \mathbf{v}_\phi} R_\phi f(\mathbf{v}) \tag{54}$$

where $\mathbf{v}_\phi = R_\phi \mathbf{v}$, and that H is independent of the orientation of the velocites, we obtain

$$R_\phi L_H \rho = -i \sum_{n=1}^{N} \left(\frac{\partial H}{\partial \mathbf{v}_{\phi n}} \cdot \frac{\partial}{\partial \mathbf{r}_n} - \frac{\partial H}{\partial \mathbf{r}_n} \cdot \frac{\partial}{\partial \mathbf{v}_{\phi n}} \right) R_\phi \rho \tag{55}$$

Since this is true for any ρ we get

$$L_H(\phi) = -i \sum_{n=1}^{N} \left(\frac{\partial H}{\partial \mathbf{v}_{\phi n}} \cdot \frac{\partial}{\partial \mathbf{r}_n} - \frac{\partial H}{\partial \mathbf{r}_n} \cdot \frac{\partial}{\partial \mathbf{v}_{\phi n}} \right) \tag{56}$$

For the velocity derivatives we have (see Eq. (45))

$$\frac{\partial}{\partial v_{\phi x}} = \cos\phi \frac{\partial}{\partial v_x} + \sin\phi \frac{\partial}{\partial v_y}$$
$$\frac{\partial}{\partial v_{\phi y}} = -\sin\phi \frac{\partial}{\partial v_x} + \cos\phi \frac{\partial}{\partial v_y} \quad (57)$$

Substituting this in Eq. (56) and separating the cosine and sine terms we obtain

$$\begin{aligned} L_H(\phi) = & - i\cos\phi \sum_{n=1}^{N} \left(\frac{\partial H}{\partial v_{nx}} \frac{\partial}{\partial r_{nx}} + \frac{\partial H}{\partial v_{ny}} \frac{\partial}{\partial r_{ny}} - \frac{\partial H}{\partial r_{nx}} \frac{\partial}{\partial v_{nx}} - \frac{\partial H}{\partial r_{ny}} \frac{\partial}{\partial v_{ny}} \right) \\ & - i\sin\phi \sum_{n=1}^{N} \left(\frac{\partial H}{\partial v_{ny}} \frac{\partial}{\partial r_{nx}} - \frac{\partial H}{\partial v_{nx}} \frac{\partial}{\partial r_{ny}} - \frac{\partial H}{\partial r_{nx}} \frac{\partial}{\partial v_{ny}} + \frac{\partial H}{\partial r_{ny}} \frac{\partial}{\partial v_{nx}} \right) \end{aligned} \quad (59)$$

or

$$L_H(\phi) = \cos(\phi) L_H + \sin(\phi) L_H^{\perp} \quad (60)$$

where $L_H^{\perp}$is the coefficient of $\sin\phi$ in Eq. (58). Note that for $\pm\pi$ rotations (velocity inversion) we have

$$L_H(\pm\pi) = -L_H \quad (61)$$

which is equivalent to Eq. (43). For $\pm\pi/2$ rotations we have

$$L_H(\pi/2) = -L_H(-\pi/2) \quad (62)$$

Now we come back to the amplitude of the $\mathcal{H}$-function (see Eq. (53))

$$\mathcal{A}_\phi(t) = \langle\langle f_1|\Lambda \exp\left[-i(L_H\cos\phi + L_H^{\perp}\sin\phi)t\right]|\rho(0)\rangle\rangle \quad (63)$$

Defining

$$\tilde{\theta}^{\perp} = \Lambda L_H^{\perp} \Lambda^{-1} \quad (64)$$

and using also the definitions in Eqs. (24), (38) we have

$$\mathcal{A}_\phi(t) = \langle\langle f_1|\exp\left[-i(\tilde{\theta}\cos\phi + \tilde{\theta}^{\perp}\sin\phi)t\right]|\tilde{\rho}(0)\rangle\rangle \quad (65)$$

With no rotation ($\phi = 0$) we have only the operator $\tilde{\theta}$ in the exponential, This operator breaks time-symmetry, as discussed in Sec. 4, giving exponential decay in the positive t direction. If we perform a $\phi = \pi$ rotation, corresponding to a velocity inversion we have $\mathcal{A}_\pi(t) = \mathcal{A}_I(t)$. As discussed in Sec. 4, $\mathcal{A}$ jumps to a higher value, the jump increasing exponentially with t. Now, if we perform a $\phi = \pi/2$ rotation we expect that the change of $\mathcal{A}$ will be the same as if we perform a $\phi = -\pi/2$ rotation, because nothing in the Hamiltonian H or the Λ transformation makes a distinction between the sense of the rotations. This implies that in contrast to $\tilde{\theta}$, the operator $\tilde{\theta}^\perp$ cannot be a dissipative operator breaking time-symmetry. If it were, then $\mathcal{A}_{\pi/2}(t)$ would contain terms growing exponentially in the future t direction while $\mathcal{A}_{-\pi/2}(t)$ would contain terms decaying exponentially, or viceversa (see Eq. (62)). We would have an asymmetry between clockwise and counter-clockwise velocity rotations. So, the function $\mathcal{A}_{\pm\pi/2}(t)$ should only contain time invariant components plus oscillating components. The spectrum of frequencies of the oscillating components is continuous, so by the Riemann-Lebesgue theorem, the oscillating components added together will give a contribution decreasing as an inverse power of t after a time scale t_c.

$$\mathcal{A}_{\pi/2}(t) \approx \mathcal{A}_{\pi/2}(0), \quad \text{for}\, t \gg t_c \tag{66}$$

For weak coupling one can estimate that $t_c \sim 1/(\lambda c^{1/2})$. Eq. (66) means that for $t \gg t_c$ we can neglect the contributions coming from the operator $\tilde{\theta}^\perp$. Thus we have

$$\mathcal{A}_\phi(t) \approx \langle\langle f_1 | \exp\left[-i(\tilde{\theta}\cos\phi)t\right] |\tilde{\rho}(0)\rangle\rangle \tag{67}$$

The $\mathcal{H}$-function is then even with respect to the angle of rotation ϕ. The $\mathcal{H}$-function jump increases with ϕ from $\phi = 0$ (no jump) to $\phi = \pi$ (maximum jump). For $\phi = \pi/2$ we have

$$\mathcal{H}_{\pi/2}(t) \approx \mathcal{H}(0) \tag{68}$$

In other words, a $\pi/2$ rotation resets the $\mathcal{H}$-function to its initial value. We have a complete rejuvenation. For $0 < \phi < \pi/2$ we have partial rejuventations, that is, the system becomes"younger" (more ordered) but not as young as it was at $t = 0$. For $\pi/2 < \phi < \pi$ we have an "over-rejuvenation:" the system becomes younger than it was at $t = 0$ due to the presence of anomalous correlations (see Fig. 2).

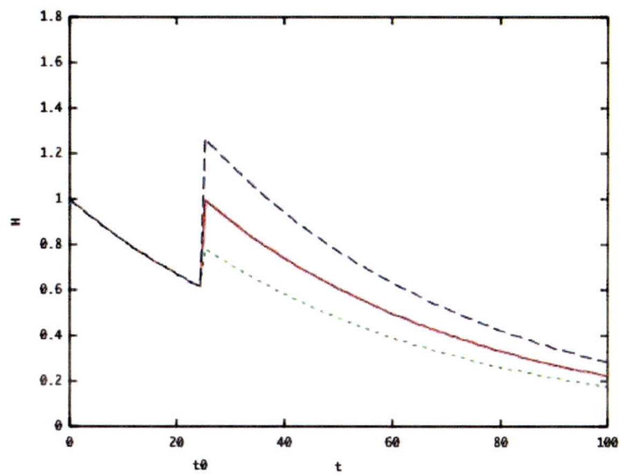

Figure 2: Schematic plot of the Lyapounov function $\mathcal{H}(t)$, showing the effects of velocity rotation at a given timet_0. Velocity rotations create correlations and cause $\mathcal{H}$ to jump up. The dashed line corresponds to a rotation by an angle $\pi/2 < \phi < \pi$ giving "over-rejuvenation", while the solid line corresponds to a $\pi/2$ rotation reseting $\mathcal{H}(t)$ to its initial value and the dotted line corresponds to a $0 < \phi < \pi/2$ rotation giving a partial rejuvenation.

6 Comparison with the twin effect

Now we come to our analogy with the twin effect. Suppose we can "sit" on one of the particles (particle 1). After a collision with another particle (particle 2), we see particle 2 move away from particle 1 with some velocity g. Let us say that the distance between the particles at the moment of the rotation was r. After a velocity rotation, the distance between the particles will change at the rate

$$\dot{r}(t) = g\frac{r\cos\phi + gt}{\sqrt{(r\cos\phi + gt)^2 + r^2\sin^2\phi}} \tag{69}$$

The distance between the particles will continue to increase if $\phi \leq \pi/2$, but it will decrease if $\phi > \pi/2$. The minimum distance between the particles will be

$$r_{min} = r\sin\phi < r, \quad \text{for}\, \pi/2 < \phi \leq \pi \tag{70}$$

As we have seen, it is precisely for this range of angles that we have an "over-rejuvenation" of the system, that is, after the rotation the system will become younger than it was at $t = 0$. The closer the particles come together as a result of the rotation, the more the system is rejuvenated. For angles $0 \leq \phi \leq \pi/2$ the rate at which the particles move away is decreased, and we have a partial rejuvenation (note that $|\dot{r}(t)| < g$ for any angle ϕ). All this is reminiscent of the twin effect, replacing particles 1 and 2 by the twins. In the twin effect, the travelling twin becomes younger than the twin at rest, as a consequence of acceleration, which is necessary to bring the twins together. The larger the acceleration, the larger the differences in ages. A small acceleration, not strong enough to bring the twins together would still lead to a small effect, slowing slightly the aging of the moving twin. Of course, this is a relativistic effect due to dilation of time intervals. So there are similarities with the twin effect, but there are also differences. In contrast to the twin effect, the situation we have considered is non-relativistic, and the rejuvenation is a collective effect involving all the particles of the system (not only particles 1 and 2). It is due to the injection of correlations among the particles, which turns post-collisional correlations into pre-collisional correlations.

7 Concluding Remarks

External forces acting on a system, leading to acceleration, can have an effect on the entropy of the system. This effect has a relativistic

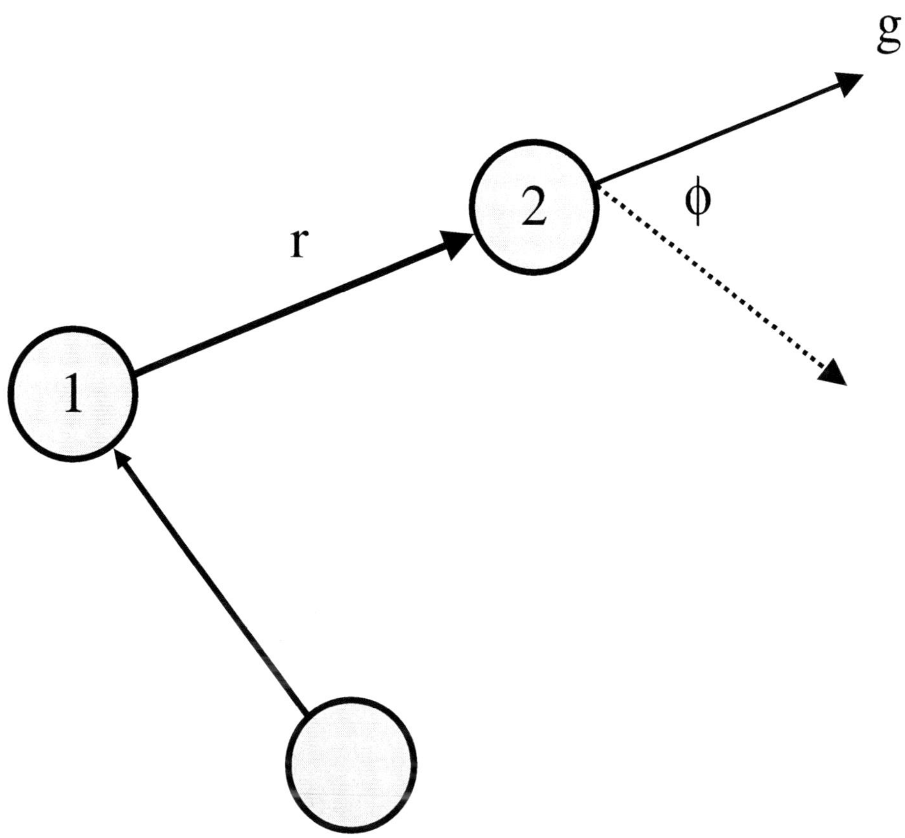

Figure 3: After colliding with particle 1, particle 2 moves away from particle 1 with a speed g. When they are separated a distance r, we perform a velocity rotation by an angle ϕ. This process, applied to all collisions inside the gas, leads to a "rejuvenation" of the system, due to the creation of new correlations among the particles. In the twin effect, an acceleration on the moving twin (analogous to particle 2) slows down his aging, as compared to the twin at rest. This effect is due to relativistic time dilation.

component, as in the twin paradox, as well as non-relativistic components, an example of which we have discussed in this paper. There is also a distinction between global forces, leading to an overall acceleration (again as in the twin paradox) and local forces, depending on the state of each particle, as we considered here.

Acknowledgments

We thank Dr. E. Karpov and Dr. T. Petrosky for helpful comments and suggestions. We acknowledge the International Solvay Institutes for Physics and Chemistry, the Engineering Research Program of the Office of Basic Energy Sciences at the U.S. Department of Energy, Grant No DE-FG03-94ER14465, the Robert A. Welch Foundation Grant F-0365, and the European Commission Project HPHA-CT-2001-40002 for supporting this work.

References

[1] I. Prigogine, *Non Equilibrium Statistical Mechanics* (Wiley Interscience, 1962).

[2] I. Prigogine, C. George, F. Henin, L. Rosenfeld, *Chemica Scripta* **4**, 5 (1973).

[3] T. Petrosky, I. Prigogine and S. Tasaki, *Physica A* **173**, 175 (1991).

[4] G. Ordonez, T. Petrosky and I. Prigogine, *Phys. Rev. A* **63**, 052106 (2001).

[5] T. Petrosky, G. Ordonez and I. Prigogine, *Phys. Rev. A* **64**, 062101 (2001).

[6] M. de Haan, C. George, and F. Mayne, *Physica A* **92**, 584 (1978).

[7] I. Prigogine, *From being to becoming* (Freeman, New York, 1980).

[8] T. Petrosky and I. Prigogine, *Adv. Chem. Phys.* **99**, 1 (1997).

[9] C. George, *Physica* **39**, 251 (1968).

In: New Research on Chaos and Complexity ISBN 1-60021-275-1
Editors: F. Orsucci and N. Sala, pp. 21-46

Chapter 3

WILLIAM JAMES ON CONSCIOUSNESS, REVISITED

Walter J. Freeman
Department of Molecular & Cell Biology, LSA 142
University of California at Berkeley CA 94720-3200
TEL 510-642-4220 FAX 510-643-6791
wfreeman@socrates.berkeley.edu

ABSTRACT

According to the behavioral theory of pragmatism described most effectively by William James in collaboration with Charles Peirce and John Dewey, knowledge about the world is gained through intentional action followed by learning. In terms of neurodynamics, when the intending of an act comes to awareness through reafference, it is perceived as a cause. When the consequences of an act come to awareness through exteroception and proprioception, they are perceived as effects. These become the cause of a new act. Cycles of such states of awareness comprise consciousness, which can grow in complexity to include self-awareness. Intentional acts do not require awareness, whereas voluntary acts require self-awareness. Awareness of the action-perception cycle provides the cognitive metaphor of linear causality as agency. Humans apply this metaphor to objects and events in the world in order to predict and control them, and to assign social responsibility. Thus linear causality is the bedrock of social contracts and technology.

Complex material systems with distributed nonlinear feedback, such as brains and the activities of their neural and behavioral substrates, cannot be explained by linear causality. They can be said to operate by circular causality without agency. The nature of self-control is described by breaking the circle into a forward limb, the intentional self, and a feedback limb, awareness of the self and its actions. The two limbs are realized through hierarchically stratified kinds of neural activity. Actions are governed by the microscopic neural activity of cortical and subcortical components in the brain that is self-organized into mesoscopic wave packets. The wave packets form by state transitions that resemble phase transitions between vapor and liquid. The cloud of action potentials driven by a stimulus condenses into an ordered state that gives the category of the stimulus. Awareness supervenes as a macroscopic ordering state that defers action until the self-organizing mesoscopic process has reached closure in reflective prediction. Agency, which is removed from the causal hierarchy by the appeal to circularity, re-appears as a metaphor by which objects and events in the world are anthropomorphized and assigned the human property of causation, so that they can be assimilated as subject to the possibility of observer control.

Key words: causality, consciousness, intentionality, nonlinear dynamics, reafference

1. INTRODUCTION

Within a single generation of the publication by Charles Darwin of "The Origin of Species", the basic concepts of evolution had been grasped and put to use by William James. He wrote: "*A priori* analysis of both brain and conscious action shows us that if the latter were efficacious it would, by its selective emphasis, make amends for the indeterminacy of the former; whilst the study *à posteriori* of the distribution of consciousness shows it to be exactly such as we might expect in an organ added for the sake of steering a nervous system grown too complex to regulate itself" (James 1879, p. 18). This is classic James: elegant, urbane, a bit fey, and precisely on target, as far as he went. But, did he go far enough? In my view, he did not. There are several shortcomings. Firstly, he proposed the addition of a new part to the brain for the addition of consciousness. We have no evidence that consciousness resides in or operates from any newly added part of the human brain, including those for language. Secondly, his definition finessed the questions whether language is necessary for consciousness, and, if not, whether animals evolved the necessary brain part, and therefore consciousness, early in phylogenetic evolution. Thirdly, he gave no indication of how consciousness might execute its steering function. Some kind of control is modeled by cybernetics, a term that Norbert Wiener coined from the Greek word for "steersman", but even to the present there is no widely accepted explanation of the nature and role of consciousness. Fourthly and more generally, he assigned a causal role to consciousness, even though he allowed (James 1890): "The word 'cause' is ... an altar to an unknown god."

What he did do was to raise and answer the question whether consciousness had a biological basis that could be selected for in the race for the survival of the fittest. He disposed of alternative views that consciousness was an epiphenomenal appendage, which was produced by the brain but which had no role in behavior, and that consciousness was an endowment from God by which humans might come to know the Almighty. He stated clearly that consciousness is known through experience of the activities of one's own body, and by observation of the bodies of others. He laid the foundation for the biological study of the properties of consciousness and of its roles in the genesis and regulation of behaviors. These properties are fair targets for experimental analyses and modeling, unlike the questions whether it arises from the soul (Eccles, 1994), or from panpsychic properties of matter (Whitehead, 1938; Penrose, 1994; Chalmers, 1996), or as a necessary but unexplained and inexplicable accompaniment of brain operations (Searle, 1992; Dennett, 1991; Crick, 1994). In the way James phrased his conception, the pertinent questions are — however it arises and is experienced — how and in what senses does consciousness cause the functions of brains and bodies, and how do brain and bodily functions cause it? How do actions cause perceptions? How do perceptions cause awareness? How do states of awareness cause actions? How can the action potentials of neurons cause consciousness, and how can consciousness shape the patterns of neural firing?

2. The Typology of Causality

Analysis of causality is an essential step to understand consciousness, because the forms that answers to these questions take, and even whether answers can exist, depend on the choice among meanings that are assigned to "cause": (a) to make, move and modulate (an agency in linear causality); (b) to explain rationalize and assign credit or blame (comprehension in circular causality without agency); or (c) to flow in parallel as a meaningful experience, by-product, or epiphenomenon (noncausal interrelations as in predictors, statistical "risk factors" or Leibnizian monads). The troublesome and problematic nature of "causes" is reflected in the variety of synonyms that have been proposed over the centuries: "dispositions" by Thomas Aquinas (1272); "tendencies" by John Stuart

Mill (1843); "anomalous monism" by David Davidson (1980); "propensities" by Karl Popper (1982); and "capacities" by Nancy Cartwright (1989).

The prior question I raise here is, why is it that we seek for an explanation of consciousness, by which it is both an effect of neural activity and a cause of behavior? In other words, what are the properties of the neural mechanisms of human thought that lead us to phrase questions in just this way? Obviously there are many answers available to us, but there is no agreement on what the basis might be for finding human satisfaction in answers that invoke causality. My aim here is to show why we humans are addicted to causality.

Linear causality is exemplified in stimulus-response determinism. A stimulus (S) initiates a chain of events including activation of receptors, transmission by serial synapses to cortex, integration with memory, selection of a motor pattern descending transmission to motor neurons, and activation of muscles. At one or more nodes along the chain awareness occurs, and meaning and emotion are attached to the response (R). Temporal sequencing is crucial; no effect can precede or occur simultaneously with its cause. At some instant each effect becomes a cause. This step is inherently problematic, because awareness cannot be defined at points in time. The demonstration of causal invariance must be based on repetition of trials, in which universal time is segmented. The time line for each observation is re-initiated at zero in observer time, and S-R pairs are collected. Some form of generalization is used over the pairs, and various forms of abstraction are used to control and exclude extraneous factors and correlations in the attempt to define true agencies.

Noncausal relations are described by statistical models, differential equations, phase portraits, and so on, in which time may be implicit and/or reversible. Once the constructions are completed by the calculation of risk factors and degrees of certainty from distributions of observed events and objects, the assignment of causation is optional. In describing brain functions by these techniques, consciousness is treated as irrelevant, epiphenomenal, or unscientific and of little further interest (Dennett 1991; Crick 1994).

Circular causality defies simple summary. My approach in this essay is to explain it at three levels of brain function: the macroscopic level of brain body and mind in relation to behavior; the mesoscopic level of neuron populations within the brain; and the microscopic level at which individual neurons act in concert to create populations. These concepts can be applied to animal consciousness, on the premise that the structures and activities of brains and bodies are comparable to those of humans over a broad variety of animals. The hypothesis is that the elementary properties of consciousness have emerged and are manifested in even the simplest of extant vertebrates. Structural and functional complexity of mind, brain and body increased with the evolution of brains into higher mammals. There were quantum leaps in complexity upon the addition of new parts subserving language and other social functions, but the dynamics of intentionality in brains was and remains couched in neural operations that construct goal-oriented behavior, for which language is neither necessary nor sufficient. The brains of invertebrates are not hereby consigned to mindless machines, because cuttlefish and bees appear to have the capacity for play, but their brains are sufficiently different topologically from those of vertebrates as to make inclusion too difficult for present purposes.

3. Level 3 - Macroscopic: The Circular Causality of Intentionality

An elementary process requiring the dynamic interaction between brain, body and world in all animals is an act of observation. This is not a passive receipt of information from the world, as expressed linear causality. It is the culmination of purposive action by which an animal directs its sense organs toward a selected aspect of the world and abstracts, generalizes, and learns from the resulting sensory stimuli. This principle was the starting point for Charles Peirce and William James in the development of pragmatism (James 1893; Menand 2001). Each such act requires a prior state of readiness that expresses the existence in the actor of a goal, a preparation for motor action by positioning the sense organs and selectively sensitizing the sensory cortices. Before stimulus arrival their excitability has already been shaped by the past experience that is relevant to the goal and the expectancy of stimuli. A concept that can serve as a principle by which to assemble and interrelate these multiple facets is intentionality. Aquinas (1272) introduced this concept in his program to Christianize Aristotelian doctrine. He conceived it on the basis of the fundamental integrity of the soul, mind and body of the individual, and the power of the individual to know God by taking action into the world ("stretching forth") and suffering the consequences. Descartes and Kant deliberately carried out revolutions against Thomist doctrine. They replaced intentionality with the concept of representationalism, in which forms and information come from the world and are transformed into images that are interpreted according to the laws of mathematics, logic and reason. Brentano (1889) resurrected intentionality but only to denote the relation between mental representations and the objects and events being represented, thus reinforcing Descartes' subject-object dichotomy.

The properties of Thomist intentionality are (a) its intent, directedness toward some future state or goal; (b) its unity; and (c) its wholeness in the integration of a life-long remembrance of experiences (Freeman 1995). (a) Intent comprises the endogenous initiation, construction and direction of behavior into the world, combined with changing the self by learning in accordance with the perceived consequences of the behavior. Intent originates within brains. Humans and other animals select their own goals, plan their own tactics, and choose when to begin, modify, and stop sequences of action. Humans at least are subjectively aware of themselves acting. This facet is commonly given the meaning of purpose and motivation by psychologists, because, unlike lawyers, they usually do not distinguish between intent and motive. Intent is the potential for a forthcoming action, whereas motive is the reason for the action to be taken. Intentions are biological; motives are mental.

(b) Unity appears in the combining of input from all sensory modalities into Gestalts (multisensory perceptions) in the coordination of all parts of the body, both musculoskeletal and autonomic, into adaptive, flexible, yet focused movements, and in the full weight of all past experience in the directing of each action. Subjectively, unity may appear in the awareness of self. Unity and intent find expression in modern analytic philosophy as "aboutness", meaning the way in which beliefs and thoughts symbolized by mental representations refer to objects and events in the world, whether real or imaginary (Searle 1992). The distinction between inner image and outer object invokes the dichotomy between subject and object that was not part of the originating, non-representationalist Thomist view.

(c) Wholeness is revealed by the orderly changes in the self and its behavior that constitute the development and maturation of the self through learning, within the constraints of its genes and its material, social and cultural environments. Subjectively, wholeness is revealed in the striving for the fulfillment of the potential of the self through its lifetime of change. Its root meaning is "tending", the Aristotelian view that biology is destiny. Its biological basis is seen in the process of healing of the brain and body from damage and disruption. The concept appears in the description by a 14th century surgeon LaFranchi of Milan of two forms of healing, by first intention with a clean scar, and by second intention with suppuration.

Intentionality cannot be explained by linear causality, because, under that concept, actions must be attributed solely to environmental (Skinner, 1969) and genetic determinants (Herrnstein and Murray, 1994), leaving no leeway for self-determination. Acausal theories (Hull, 1943; Grossberg, 1982) describe statistical and mathematical regularities of behavior without reference to intentionality. Circular causality explains intentionality in terms of "action-perception cycles"

(Merleau-Ponty, 1945) and affordances (Gibson 1979), in which each perception concomitantly is an outcome of a preceding action and a condition for a following action. Dewey (1914) phrased the same idea in different words; an organism does not react to a stimulus but acts into it and incorporates it. That which is perceived already exists in the perceiver, because it is posited by the action of search and is actualized in the fulfillment of expectation. The unity of the cycle is reflected in the impossibility of defining a moving instant of 'now' in subjective time, as an object is conceived under linear causality. The Cartesian distinction between subject and object does not appear, because they are joined by assimilation in a seamless flow.

4. Level 2 - Mesoscopic: The Circular Causality of Reafference

Brain scientists have known for over a century that the necessary and sufficient part of the vertebrate brain to sustain minimal intentional action as a component of intentionality, is the ventral forebrain including those parts that comprise the external shell of the phylogenetically oldest part of the forebrain the paleocortex, and the underlying nuclei such as the amygdala and the neurohumoral brain stem nuclei (Panksepp, 1998) with which the cortex is interconnected. These components suffice to support identifiable patterns of intentional behavior in animals, when all of the newer parts of the forebrain have been surgically removed (Goltz, 1892) or chemically inactivated by spreading depression (Bures et al., 1974). Intentional behavior is severely altered or lost following major damage to these parts. Phylogenetic evidence comes from observing intentional behavior in salamanders, which have the simplest of the existing vertebrate forebrains (Herrick, 1948; Roth, 1987) comprising only the limbic system. Its three cortical areas are sensory (which is predominantly the olfactory bulb), motor (the pyriform cortex), and associational (Figure 1). The latter has the primordial hippocampus interconnected with the septal, amygdaloid and striatal nuclei. It is identified in higher vertebrates as the locus of the functions of spatial orientation (the "cognitive map") and temporal orientation ("short term memory") in learning. These integrative frameworks are essential for intentional action into the world, because even the simplest actions, such as observation, searching for food, or evading predators require an animal to coordinate its position in the world with that of its prey or refuge, and to evaluate its progress during evaluation, attack or escape. These limbic structures in the medial temporal lobe appear to be the principal controllers of the neurohumoral nuclei in the hypothalamus, periacqueductal gray matter, and brain stem that are essential for the elaboration deployment and maintenance of states of readiness to act, which we identify with emotion (James, 1893; Panksepp, 1998).

The crucial question for neuroscientists is, how are the patterns of neural activity that sustain intentional behavior constructed in brains? A route to an answer is provided by studies of the electrical activity of the primary sensory cortices of animals that have been trained to identify and respond to conditioned stimuli. An answer appears in the capacity of the cortices to construct novel patterns of neural activity by virtue of their self-organizing dynamics.

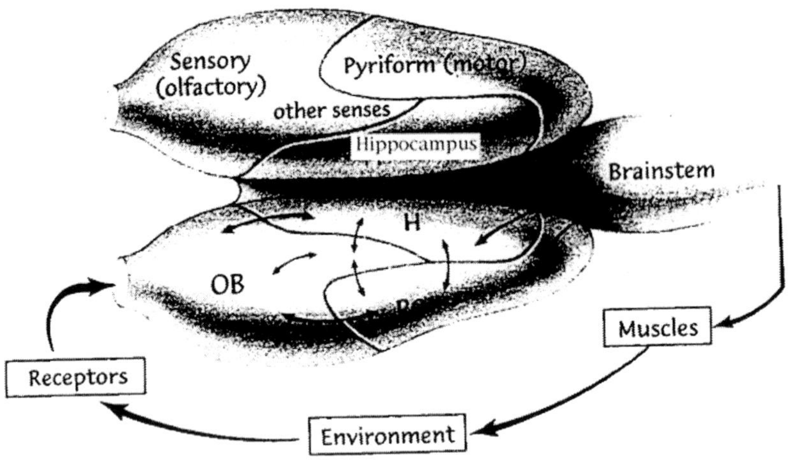

Figure 1. The schematic shows the dorsal view of the right cerebral hemisphere of the salamander (adapted from Herrick 1948). The cortical interactions are demarcated by arrows between the sensory area (olfactory bulb) with a 'transitional zone' (Tr) for all other senses and a motor area (Pir, pyriform cortex with descending connections to the corpus striatum (CS), amygdaloid (A) and septum, S), and both with the primordial hippocampus (Hip). This primitive forebrain suffices as an organ of intentionality, comprising the limbic system (Freeman 1999).

Two approaches to the study of sensory cortical dynamics are in contrast. One is based in linear causality. An experimenter identifies a neuron in sensory cortex by recording its action potential with a microelectrode, and then determines the sensory stimulus or motor action with which that neuron is most closely correlated. The pulse train of the neuron is treated as a symbol to 'represent' that stimulus as the 'feature' of an object, for example the color, contour, or motion of an eye or a nose in a face, or a ‘command’ for an action. The pathway of activation from the sensory receptor through relay nuclei to the primary sensory cortex and then beyond is described as a series of maps, in which successive representations of the stimulus are derived. The firings of the feature detector neurons must then be synchronized or 'bound' together to represent the object, such as a moving colored ball, as it is conceived by the experimenter. Neurobiologists postulate that this representation is transmitted to a higher cortex, where it is compared with representations of previous objects that are retrieved from memory storage. A solution to the 'binding problem' is still being sought (Gray, 1994; Hardcastle, 1994; Singer and Gray, 1995).

The other approach is based in circular causality. In this view the experimenter trains a subject to cooperate with him or her through use of positive or negative reinforcement, thereby inducing a state of expectancy and search for a stimulus, as the subject conceives it. When the expected stimulus arrives, the activated receptors transmit pulses to the sensory cortex, where they elicit the construction by nonlinear dynamics of a macroscopic, spatially coherent oscillatory pattern that covers the entire primary sensory cortex (Freeman 1975, 1991). Such patterns are observed by means of the electroencephalogram (EEG) from electrode arrays on any or all of the sensory cortices (Freeman 1975, 1992, 1995; Freeman and Schneider, 1978; Barrie et al., 1996; Kay and Freeman 1998). They are not seen in recordings from single neuronal action potentials, because the fraction of the variance in the typical single neuronal pulse train that is covariant with the neural mass is far too small, on the order of 0.1%.

The emergent pattern is not a representation of a stimulus, nor is it a ringing as when a bell is struck, nor a resonance as when one string of a guitar vibrates when another string does so at its natural frequency. It is a state transition that is induced by a stimulus, followed by a construction of a spatial pattern of amplitude modulation (AM) of the rapid oscillations in potential. The AM pattern is shaped by the synaptic modifications among cortical neurons from prior learning. It is also dependent on the brain stem nuclei that bathe the forebrain in neuromodulatory chemicals. It is a dynamic action pattern that creates and carries the meaning of the stimulus for the subject. It reflects the individual history, present context, and expectancy, corresponding to the unity and the wholeness of intentionality. Owing to dependence on history, the AM patterns created in each cortex are unique to each subject. The first event in neocortex upon stimulus arrival maintains information that relates to the stimulus directly, but the events thereafter reflect the category of the stimulus, its value, significance and meaning (Ohl, Scheich and Freeman 2001). This is because the mechanism of construction derives from destabilization of the cortex by input, which increases the density of excitatory interactions among neurons in the cortex, so that their activity reflects predominantly the synaptic modifications in cortex from previous learning, not the activity driven by the input (Freeman 1991; Freeman and Barrie 2000). These properties have been simulated in models both in software (Kozma and Freeman 2001) and in VLSI hardware (Principe et al. 2001). They demonstrate the difference between the passive representation of stimuli and the active engagement of the brain in the enactment of the meanings of stimuli.

The visual, auditory, somesthetic and olfactory cortices serving the distance receptors all transmit their constructions through the entorhinal cortex from whence they converge into the limbic system, where they are integrated with each other over time. Clearly they must have similar dynamics, in order that the messages be combined into Gestalts. The resultant integrated meaning is transmitted back to the cortices in the processes of selective attending, expectancy, and the prediction of future inputs (Freeman 1995; Kay and Freeman 1998; Ohl, Scheich and Freeman 2001). The same waveforms of EEG activity as those found in the sensory cortices are found in various parts of the limbic system. This similarity indicates that the limbic system also has the capacity to create its own spatiotemporal patterns of neural activity. The patterns are embedded in past experience and convergent multisensory input, and they are self-organized. The limbic system provides interconnected populations of neurons that, according to the hypothesis being proposed, generate continually the patterns of neural activity that form goals and direct behavior toward them.

EEG evidence shows that the process in the various areas of cortex occurs in discontinuous steps, like frames in motion pictures on multiple screens (Freeman 1975; Barrie, Freeman and Lenhart, 1996). Being intrinsically unstable, the limbic system continually transits across states that emerge, transmit to other parts of the brain, and then dissolve to give place to new ones. Its output controls the brain stem nuclei that serve to regulate its excitability levels, implying that it regulates its own neurohumoral context, enabling it to respond with equal facility to changes both in the body and the environment that call for arousal and adaptation or rest and recreation. Again by inference, it is the neurodynamics of the limbic system, with contributions from other parts of the forebrain such as the frontal lobes and basal ganglia, that initiates the novel and creative behaviors seen in search by trial and error.

The limbic activity patterns of directed arousal and search are sent into the motor systems of the brain stem and spinal cord (Figure 2). Simultaneously, patterns are transmitted to the primary sensory cortices, preparing them for the consequences of motor actions. This process has been called "reafference" (von Holst and Mittelstädt 1950; Freeman 1995), "corollary discharge" (Sperry 1950), and "preafference" (Kay and Freeman 1998). It compensates for the self-induced changes in sensory input that follow the actions organized by the limbic system, and it selectively sensitizes sensory systems to anticipated stimuli prior to their expected times of arrival.

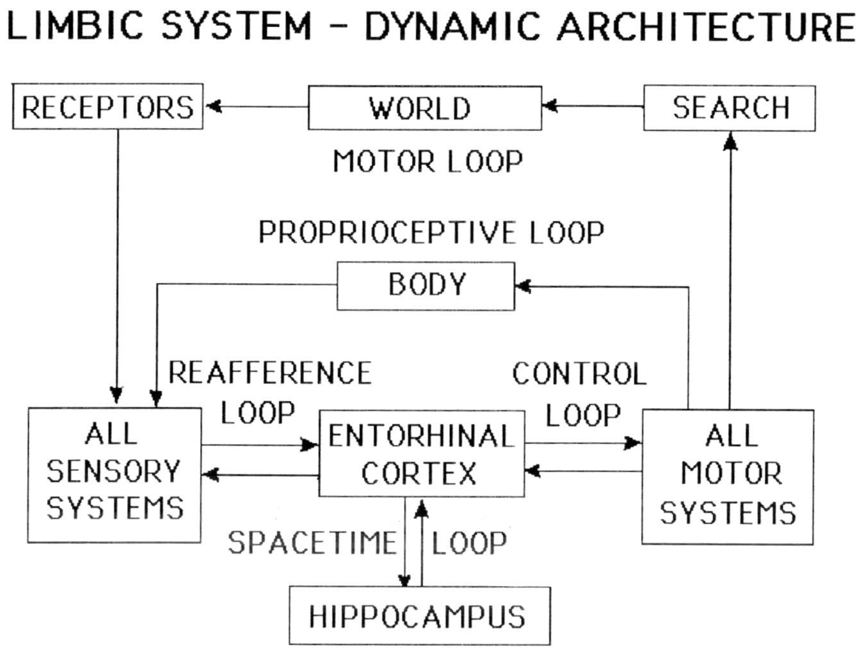

Figure 2. The limbic architecture is formed by multiple loops. The mammalian entorhinal cortex receives from and transmits to all sensory areas. It provides the main input for the hippocampus and is the main target for hippocampal output. The hypothesis is proposed that intentional action is by flow of activity around the loops that extend into the body and the world, and that awareness and consciousness are engendered by the flows within brains that, are described by circular causality.

The concept of preafference began with an observation by Helmholtz (1872) on patients with paralysis of lateral gaze, who, on trying and being unable to move an eye, reported that the visual field appeared to move in the opposite direction. He concluded that "an impulse of the will" that accompanied voluntary behavior was unmasked by the paralysis. He wrote: "These phenomena place it beyond doubt that we judge the direction of the visual axis only by the volitional act by means of which we seek to alter the position of the eyes.". J. Hughlings Jackson (1931) repeated the observation but postulated alternatively that the phenomenon was caused by "an in-going current", which was a signal from the non-paralyzed eye that moved too far in the attempt to fixate an object, and which was not a recursive signal from a "motor centre". Edward Titchener (1907) and, unfortunately, William James (1893) joined in this interpretation, thus delaying deployment of the concepts of neural feedback and re-entrant cognitive processes until late in the 20th century.

The sensory cortical constructions consist of staccato messages to the limbic system, which convey what is sought and the result of the search. After multisensory convergence, the spatiotemporal activity pattern in the limbic system is up-dated through temporal integration in the hippocampus. Accompanying sensory messages there are return up-dates from the limbic system to the sensory cortices, whereby each cortex receives input that has been integrated with the input from all others, reflecting the unity of intentionality. Everything that a human or an animal knows comes from the circular causality of action, preafference, perception and assimilation. Successive frames of self-organized activity patterns in the sensory and limbic cortices embody the cycle. This is the full program that was implicit in James' pragmatism, before the electrophysiological techniques of brain imaging made explicit the preconscious neural operations of intentionality.

5. Level 1 - Microscopic: Circular Causality Among Neurons and Neural Masses

The "state" of the brain is a description of what it is doing in some specified time period. A state transition occurs when the brain changes and does something else. For example, locomotion is a state, within which walking is a rhythmic pattern of activity that involves large parts of the brain spinal cord, muscles and bones. The entire neuromuscular system changes almost instantly with the transition to a pattern of jogging or running. Similarly, a sleeping state can be taken as a whole, or divided into a sequence of slow wave and REM stages. Transit to a waking state can occur in a fraction of a second, whereby the entire brain and body shift gears, so to speak. The state of a neuron can be described as active and firing or as silent, with sudden changes in patterns of firing constituting state transitions. Populations of neurons also have a range of states, such as slow wave, fast activity, seizure, or silence. The science of dynamics describes states and their transitions.

The most critical question to ask about a state is its degree of stability or resistance to change or dissolution. Stability is evaluated by perturbing an object or a system (Freeman 1975). For example, an egg on a flat surface is unstable, but a coffee mug is stable. A person standing on a moving bus and holding on to a railing is stable, but someone walking in the aisle is not. If a person regains his chosen posture after each perturbation no matter in which direction the displacement occurs, that state is regarded as stable, and it is said to be governed by an attractor. This is a metaphor to say that the system goes ("is attracted") to the steady state through interim transience. The range of displacement from which recovery can occur defines the basin of attraction in analogy to a ball rolling to the bottom of a bowl. If a perturbation is so strong that it causes concussion or a broken leg, so that the person cannot stand up again, then the system has been placed outside the basin of attraction, and a new state supervenes with its own attractor and basin of attraction.

Stability is always relative to the time duration of observation and the criteria for what the experimentalist chooses to observe. In the perspective of a lifetime, brains appear to be highly stable, in their numbers of neurons, their architectures and major patterns of connection, and in the patterns of behavior they produce, including the character and identity of the individual that can be recognized and followed for many years. A brain undergoes repeated state transitions from waking to sleeping and back again coming up refreshed with a good night or irritable with insomnia, but still, giving arguably the same person as the night before. But in the perspective of the short term, brains are highly unstable. Thoughts go fleeting through awareness, and the face and body scintillate with the flux of emotions. Glimpses of the internal states of neural activity reveal patterns that are more like hurricanes than the orderly march of symbols in a computer, with the difference that hurricanes don't learn. Brain states and the states of populations of neurons that interact to give brain function are highly irregular in spatial form and time course. They emerge, persist for a small fraction of a second, then disappear and are replaced by other states.

Neuroscientists aim to describe and measure these states and tell what they signify for observations of behavior and experiences with awareness. The approach of dynamics is by defining three kinds of stable state, each with its type of attractor. The simplest is the point attractor. The system is at rest unless perturbed, and it returns to rest when allowed to do so. As it relaxes to rest, it has a brief history that is lost upon convergence to rest. Examples of point attractors are neurons or neural populations that have been isolated from the brain and also the brain that is depressed into inactivity by injury or a strong anesthetic, to the point where the EEG has gone flat. A special case of a point attractor is noise. This state is observed in populations of neurons in the brain of a subject at rest, with no evidence of overt behavior or awareness. The neurons fire continually but not in concert with each other. Their pulses occur in long trains at irregular times.

Knowledge about the prior pulse trains from each neuron and those of its neighbors up to the present fails to support the prediction of when the next pulse will occur. The state of noise has continual activity with no history of how it started, and it gives only the expectation that its average amplitude and other statistical properties will persist unchanged.

A system that gives periodic behavior is said to have a limit cycle attractor. The classic example is the clock. When it is viewed in terms of its ceaseless motion it is regarded as unstable until it winds down runs out of power, and goes to a point attractor. If it resumes its regular beat after it is re-set or otherwise perturbed, it is stable as long as its power lasts. Its history is limited to one cycle, after which there is no retention of its transient approach in its basin to its attractor. Neurons and populations rarely fire periodically, and when they appear to do so, close inspection shows that the activities are usually irregular and unpredictable in detail, and when periodic activity does occur, it is likely to be either intentional, as in rhythmic drumming, or pathological, as in nystagmus and Parkinsonian tremor.

The third type of attractor gives aperiodic oscillation of the kind that is observed in recordings of EEGs and of physiological tremors. There is no one or small number of frequencies at which the system oscillates. The system behavior is therefore unpredictable, because performance can only be projected far into the future for periodic behavior. This type was first called "strange"; it is now widely known as "chaotic". The existence of this type of oscillation was known to mathematicians a century ago, but systematic study was possible only recently after the full development of digital computers. The best-known simple systems with chaotic attractors have a small number of components and a few degrees of freedom, as for example, the double-hinged pendulum. These model systems are noise-free, stationary and autonomous, meaning that they do not interact with their environments. Large and complex systems such as neurons and brains are noisy, unstable, and engaged with their environments, so the low-dimensional models do not apply. The hallmark of chaos is the capacity for rapid switching between states despite high dimensionality, plus the capacity for creating information in novel patterns. These are the properties that make chaotic dynamics interesting and applicable to behavior, even though proof is not yet possible at the present level of developments in mathematics.

The discovery of chaos has profound implications for the study of brain function (Skarda and Freeman 1987). A dynamic system has a collection of attractors, each with its basin which forms an 'attractor landscape' with all three types. The state of the system can jump from one to another in an itinerant trajectory (Tsuda 2001). Capture by a point or limit cycle attractor wipes clean the history upon asymptotic convergence, but capture in a chaotic basin engenders continual aperiodic activity, thereby creating novel, unpredictable patterns that retain its recent history. Although the trajectory is not predictable, the statistical properties such as the mean and standard deviation of the state variables of the system serve as measures of its steady state. Chaotic fluctuations carry the system endlessly around in the basin. However, if energy is fed into the system so that the fluctuations increase in amplitude, or if the landscape of the system is changed so that the basin shrinks or flattens, a microscopic fluctuation can carry the trajectory across the boundary between basins to another attractor.

In the dynamic space of each sensory cortex there are multiple chaotic attractors with basins corresponding to previously learned classes of stimuli, including those for learned background stimulus configurations. These multiple basins and attractors constitute an attractor landscape. Chaotic prestimulus states of expectancy establish the sensitivity of the cortex by warping the landscape, so that a small number of sensory action potentials driven by an expected stimulus (the "figure"), accompanied by a large number of action potentials from irrelevant stimuli (the "background"), can carry the cortical trajectory into the basin of an appropriate attractor. Circular causality enters in the following way. The state of a neural population in an area of cortex is a macroscopic event that arises through the interactions of the microscopic activity of the neurons comprising the neuropil. The global state is upwardly generated by the microscopic neurons, and simultaneously the global state downwardly organizes ("enslaves" — Haken 1983) the activities of the individual neurons.

Each cortical state transition requires this circularity. It is preceded by a conjunction of antecedents. A stimulus is sought by the limbic brain through orientation of the sensory receptors in sniffing, looking, and listening. The landscape of the basins of attraction is shaped by limbic preafference, which facilitates access to an attractor by expanding its basin for the reception of a desired class of stimuli as with Voronoi tesselations. Preafference provides the ambient context by multisensory divergence from the limbic system. The web of synaptic connections modified by prior learning maintains the basins and attractors. Pre-existing chaotic fluctuations are enhanced by input, forcing the selection of a new macroscopic state that then engulfs the stimulus-driven microscopic activity.

There are two reasons that all the sensory systems (visual, auditory, somatic and olfactory) operate this way. First, the brain faces the infinite complexity of the world with a finite capacity for understanding. The solution for the brain is to create its own information in the form of hypotheses, which it tests by acting into the environment. In olfaction for example, a significant odorant may consist of a few molecules mixed in a rich and powerful background of undefined substances, and the odorant the brain seeks may be continually changing in age, temperature, and concentration. Each sniff in a succession with the same chemical activates a different subset of equivalent olfactory receptors, so the microscopic input is unpredictable and unknowable in detail. Detection and tracking require abstraction to an invariant pattern over trials. The attractor provides the abstraction, and the basin provides the generalization over equivalent receptors. The attractor determines the response, not the particular stimulus. Unlike the view proposed by stimulus-response reflex determinism, the dynamics gives no linear chain of cause and effect from stimulus to response that can lead to the necessity of environmental determinism. The second reason that all sensory systems operate in the same way is the requirement that the neural outputs of all sensory systems have the same basic form, so that they can be combined into Gestalts, as they are converged by integration and extension over time in the entorhinal-hippocampal system (Freeman. 1995).

6. Circular Causality in Awareness

Circular causality underlies all state transitions in sensory cortices and the olfactory bulb, when fluctuations in microscopic activity exceed a certain threshold, such that a new macroscopic oscillation emerges that forces cooperation on the very neurons that have brought the pattern into being. EEG measurements show that multiple patterns self-organize independently in overlapping time frames in the several sensory and limbic cortices. The patterns emerge and coexist briefly with stimulus-driven activity in specialized areas of the neocortex that receive the projections of sensory pathways. On transmission of cortical output, the created patterns are broadcast widely, whereas the stimulus-driven activity, having done its work, is deleted. The overlapped cortical output patterns are combined into a hemisphere-wide pattern by the neocortex, which structurally is an undivided sheet of neuropil in each hemisphere .

Circular causality can serve as the framework for explaining the operation of awareness in the following way. The multimodal macroscopic patterns converge simultaneously into the limbic system, and the results of integration over time and space are simultaneously returned to all of the sensory systems. Here comes into play James' "... organ added for the sake of steering a nervous system grown too complex to regulate itself ...", though it is not another organ but, instead, another hierarchical level in brain function: a hemisphere-wide attractor, for which the local mesoscopic activity patterns are the components. The forward limb of the loop provides the bursts of oscillations converging into the limbic system that destabilize it to form new patterns. The feedback limb incorporates the limbic and sensory cortical patterns into a global activity pattern or order parameter that enslaves all of the components (Haken 1983). The enslavement enhances the coherence among all of them, which dampens the chaotic fluctuation instead of enhancing it, as the receptor input does in the sensory cortices.

A global operator of this kind must exist, for the following reason. The synthesis of sense data first into cortical wave packets and then into a multimodal packet takes time. After a Gestalt has been achieved through embedding in past experience, a decision is required as to what the organism is to do next. This also takes time (Libet 1994) for an evolutionary trajectory through a sequence of attractors constituting the attractor landscape of possible goals and actions (Tsuda 2001). The triggering of a state transition in the motor system may occur at any time, if the fluctuations in its multiple inputs are large enough, thereby terminating the search trajectory. In some emergent behavioral situations an early response is most effective: action without reflection. In complex situations with unclear ramifications into the future, precipitate action may lead to disastrous consequences. More generally, the forebrain appears to have developed in phylogenetic evolution as an organ taking advantage of the time provided by distance receptors for the interpretation of raw sense data. The quenching function of a global operator to delay decision and action can be seen as a necessary complement on the motor side, to prevent premature closure of the process of constructing and evaluating alternative courses of possible action.

Action without the deferral that is implicit in awareness can be found in so-called 'automatic' sequences of action in the performance of familiar complex routines such as driving a car, and in thoughtless and potentially self-destructive actions that Davidson (1980) described as "incontinent". Actions can "flow" without cautionary awareness. Implicit cognition is continuous, and it is simply unmasked in the conditions that lead to 'blindsight'. In this view, emotion is defined as the impetus for action, more specifically, for impending action. Its degree is proportional to the amplitudes of the chaotic fluctuations in the limbic system, which appear in the modulation depths of the carrier waves of limbic neural activity patterns (Lesse 1957). In accordance with the James-Lange theory of emotion (James 1893), it is experienced through awareness of the activation of the autonomic nervous system in preparation for and support of overt action, as described by Cannon (1939). It is observed in the patterns of behavior that social animals have acquired through evolution (Darwin 1872). Emotion is not in opposition to reason. Behaviors that are seen as irrational and incontinent result from premature escape of the chaotic fluctuations from the leavening and smoothing of the awareness operator. The most intensely emotional behavior, as it is experienced in artistic creation scientific discovery, and religious awe, occurs as the intensity of restraining awareness rises in concert with the strength of the fluctuations (Freeman 1995). As with all other difficult human endeavors, self-control is achieved through long and arduous practice. As James correctly surmised (James 1893), the contents of moments of awareness are severely limited, giving rise to his metaphor of a "penumbra" around a moving spot of light to describe the seamless texture of intertwined meanings, knit together by the awareness operator, with continuous influence of contents well outside the penumbra constituting the unconscious.

Evidence for the existence of the postulated global operator is found in the high level of covariance in the EEGs simultaneously recorded from the bulb and the visual, auditory, somatic and limbic (entorhinal) cortices of animals, and from the scalp of humans (Lehmann and Michel 1990; Miltner et al. 1999; Rodriguez et al. 1999; Müller 2000; Csibra et al. 2000; Tallon-Baudry et al. 1998; Haig et al. 2000). The magnitude of the shared activity can be measured in limited circumstances by the largest component in principle components analysis (PCA). Even though the waveforms of the several sites vary independently and unpredictably, the first component has 50-70% of the total variance (Smart et al., 1997; Gaál and Freeman 1998; Freeman, Gaál and Jörsten, 2003). These levels are lower than those found within each area of 90-98% (Barrie, Freeman and Lenhart, 1996), but they are far greater than can be accounted for by any of a variety of statistical artifacts or sources of correlation such as volume conduction pacemaker driving, or contamination by the reference lead in monopolar recording. The high level of coherence holds for all parts of the EEG spectrum and for aperiodic as well as near-periodic waves. The maximal coherence appears to have zero phase lag over distances up to several centimeters between recording sites and even between hemispheres (Singer and Gray, 1995). Attempts are being made to model the observed zero time lag among the structures by cancellation of delays in bidirectional feedback transmission (König and Schillen 1991; Traub et al. 1996; Roelfsma et al., 1997).

7. Consciousness Viewed as a System Parameter Controlling Chaos

An unequivocal choice can be made now between the three meanings of causality proposed in the Introduction. Consciousness and neural activity are not acausal processes operating in parallel, nor does either make or move the other as an agency in temporal sequences. Circular causality is a form of explanation that can be applied at several hierarchical levels without recourse to agency. This formulation provides the sense or feeling of necessity that is essential for human satisfaction, by addressing the elemental experience of cause and effect in acts of observation, even though logically it is very different from linear causality in all aspects of temporal order, spatial contiguity, and invariant reproducibility. The phrase is a cognitive metaphor. It lacks the attribute of agency, unless and until the loop is broken into the forward (microscopic) limb and the recurrent (macroscopic) limb, in which case the agency that is so compelling in linear causality can be re-introduced. This move acquiesces to the needs of the human observers, who use it in order to seek the quale of certainty in causation by studies of dynamic events and processes in the world.

I propose that the globally coherent brain activity, which can be recorded from scalp sensors noninvasively in human subjects, and which can be interpreted as an order parameter, may be an objective correlate of awareness through preafference, comprising expectation and attention, which are based in prior proprioceptive and exteroceptive feedback of the sensory consequences of previous actions, after they have undergone limbic integration to form Gestalts, and in the goals that are emergent in the limbic system. In this view, awareness is basically akin to the intervening state variable in a homeostatic mechanism, which is a physical quantity, a dynamic operator, and the carrier of influence from the past into the future that supports optimizing the relation between a desired set point and an existing state. The content of the awareness operator may be found in the spatial pattern of amplitude modulation of the shared waveform component, which is comparable to the amplitude modulation (AM patterns) of the carrier waves in the primary sensory receiving areas.

What is most remarkable about this operator is that it appears to be antithetical to initiating action. It provides a pervasive neuronal bias that does not induce state transitions, but defers them by quenching local fluctuations (Prigogine 1980). It alters the attractor landscapes of the lower order interactive masses of neurons that it enslaves. In the dynamic view, intervention by states of awareness in the process of consciousness organizes the attractor landscape of the motor systems, prior to the instant of its next state transition: the moment of choosing in the limbo of indecision when the global dynamic brain activity pattern is increasing its complexity and fine-tuning the guidance of overt action. This state of uncertainty and unreadiness to act may last a fraction of a second, a minute, a week, or a lifetime. Then when a contemplated act occurs, awareness follows the onset of the act and does not precede it.

In that hesitancy, between the last act and the next, comes the window of opportunity, when the breaking of symmetry in the next limbic state transition will make apparent what has been chosen. The observer of the self intervenes by awareness that organizes the attractor landscape, just before the instant of the next transition. The causal technology of self-control is familiar to everyone: hold off fear and anger; defer closure; avoid temptation; take time to study; read and reflect on an opportunity with its meaning and consequences; take the long view as it has been inculcated in the educational process. According to Mill (1843): "We cannot, indeed, directly will to be different from what we are; but neither did those who are supposed to have formed our characters directly will that we should be what we are. Their will had no direct power except over their own actions. ... We are exactly as capable of making our own character, *if we will*, as others are of making it for us" (p. 550).

There are numerous unsolved problems with this hypothesis. Although strong advances are being made in analyzing the dynamics of the limbic system and its centerpieces, the entorhinal cortex and hippocampus (Boeijinga and Lopes da Silva, 1988; O'Keefe and Nadel, 1978; Rolls et al., 1989; McNaughton 1993; Wilson and McNaughton 1993; Buzsaki, 1996; Eichenbaum, 1997; Traub et al., 1996), their self-organized spatial patterns, their precise intentional contents and their mechanisms of formation in relation to intentional action are still unknown. The prepyriform cortex to which the bulb transmits is strongly driven by its input, and it lacks evidence for self-organizing state transitions (Freeman and Barrie 2000) comparable to those of the sensory cortices. Whether the hippocampus has those capabilities or is likewise a driven structure is unknown. The neural mechanisms by which the entire neocortical neuropil in each hemisphere maintains spatially coherent activity over a broad spectrum with nearly zero time lag are still undefined. In order to establish the significance of this coherent activity for behavior, it will be necessary to find and classify the mental correlates of the spatial patterns of brain electrical activity. If those correlates are meanings (Freeman, 2003), then the subjects must be asked to make representations of the meanings in their minds, in order to communicate them to observers. If the subjects are animals, their representations of meanings are restricted to gestures and goal-directed actions. If the subjects are humans, they can speak, write, draw, and make music. Given the new techniques for brain imaging now available to us, knowledge of human brain function may well be within the present reach of neurodynamics, in particular the Jamesian operator that has evolved to match and manage the complexity of our brains.

8. CONCLUSION

Consciousness in the neurodynamic view is a global internal state variable and self-regulating operator acting in a sequence of momentary states of awareness. It is essential for incorporating each new frame of awareness in the life history of an individual, which is the wholeness of intentionality. Its regulatory role is minimally comparable to that of the operator in a thermostat, that instantiates the difference between the sensed temperature and a set point, and that initiates corrective action by turning a heater on or off. The machine state variable has little history and no capacities for learning or determining its own set point, but the principle is the same: the internal state is a form of energy, an operator, a predictor of the future, and a carrier of information that is available to the system as a whole. A thermostat is a prototype, an evolutionary precursor, not to be confused with awareness, any more than tropism in plants and bacteria is to be confused with intentionality. In the Jamesian framework consciousness is the utilitarian organizer of whatever working parts the brain can provide at its present level of evolution or devolution, to use the terms of J. Hughlings Jackson (1931). In humans, the operations and informational contents of the global state variable, which are sensations, images, feelings, thoughts and beliefs, constitute the experience of cause and effect.

To deny this comparability and assert that humans are not machines is to miss the point. Two things distinguish humans from all other beings. One is the form and function of the human body, including the brain, which has been given to us by three billion years of biological evolution. The other is the heritage given to us by two million years of cultural evolution. Our mental attributes have been characterized for millennia as the soul or spirit or consciousness that makes us not-machines. The uniqueness of the human condition is not thereby explained, but the biological foundation forged by James is enhanced by the concept of circular causality. It provides a tool for intervention when something has gone wrong, because the circle can be broken into forward and feedback limbs. Each of the limbs, physical and mental, can be explained by linear causality, which can enable us to understand where and how we might intervene. The only error would be to assign causal agency to the parts of the brain instead of to ourselves as physicians and psychiatrists, as we act in the belief of our efficacy as causal agents, yet still with the humility expressed in the epitaph of Ambroise Paré, 16th century French surgeon extraordinary: "Je le pansay, Dieu le guarit" ("I bound his wounds, God healed him").

ACKNOWLEDGMENTS

This research was supported by grants from the National Institutes of Health MH-06686 and the Office of Naval Research N00014-93-1-0938. The essay appeared in an earlier version in the **Journal of Consciousness Studies** 6 Nov/Dec: 143-172, 1999 and in book form in "Reclaiming Cognition" edited by R. Núñez and W. J. Freeman.

REFERENCES

Aquinas, St. Thomas (1272) *The Summa Theologica*. Translated by Fathers of the English Dominican Province. Revised by Daniel J Sullivan. Published by William Benton as Volume 19 in the Great Books Series. Chicago: Encyclopedia Britannica, Inc., 1952.

Barrie JM, Freeman WJ & Lenhart M (1996) Modulation by discriminative training of spatial patterns of gamma EEG amplitude and phase in neocortex of rabbits. *J. Neurophysiol.* 76: 520-539.

Boeijinga, P.H. & Lopes da Silva, F.H. (1988) Differential distribution of beta and theta EEG activity in the entorhinal cortex of the cat. *Brain Res.* 448: 272-286, 1988.

Brentano FC (1889) *The Origin of our Knowledge of Right and Wrong*. Chisolm RM and Schneewind EH (trans.). New York: Humanities Press (1969).

Bures J, Buresová O, Krivánek J (1974) *The Mechanism and Applications of Leão's Spreading Depression of Electroencephalographic Activity*. New York: Academic Press.

Buzsaki G. (1996) The hippocampal-neocortical dialogue. *Cerebral Cortex* 6: 81-92.

Cannon WB (1939) *The Wisdom of the Body*. New York: WW Norton

Cartwright N (1989) *Nature's Capacities and their Measurement*. Oxford UK: Clarendon Press.

Chalmers DJ (1996) *The Conscious Mind. In Search of a Fundamental Theory*. New York: Oxford University Press.

Crick F (1994) *The Astonishing Hypothesis: The Scientific Search for the Soul*. New York: Scribner.

Csibra G, Davis G, Spratling MW, Johnson MH (2000) Gamma oscillations and object processing in the infant brain. *Science* 290: 1582-1585.

Darwin C (1872) *The Expression of the Emotions in Man and Animals*. London UK: J. Murray.

Davidson D (1980) Actions, reasons, and causes. In: *Essays on Actions & Events.* Oxford UK: Clarendon Press.

Dennett DH (1991) *Consciousness Explained.* Boston: Little, Brown.

Dewey J (1914) Psychological doctrine in philosophical teaching. *J. Philosophy* 11: 505-512.

Eccles JC (1994) *How the Self Controls Its Brain.* Berlin: Springer-Verlag.

Eichenbaum H. (1997) How does the brain organize memories? *Science* 277: 330-332.

Foucault M (1976) *The History of Sexuality: Vol. 1. An Introduction* (R Hurley, trans.). New York: Random House (1980).

Freeman WJ (1975) *Mass Action in the Nervous System.* New York: Academic

Freeman WJ (1984) Premises in neurophysiological studies of learning. Ch 13 in Lynch G, McGaugh JL, Weinberger NM (eds.) *Neurobiology of Learning and Memory.* New York: Guilford Press.

Freeman WJ (1991) The physiology of perception. *Scientific American* 264: 78-85.

Freeman WJ (1992) Tutorial in Neurobiology: From Single Neurons to Brain Chaos. *Int. J. Bifurc. Chaos* 2: 451-482.

Freeman WJ (1995) *Societies of Brains. A Study in the Neuroscience of Love and Hate.* Mahwah NJ: Lawrence Erlbaum Associates.

Freeman WJ (1999) *How Brains Make Up Their Minds.* London: Weidenfeld & Nicolson.

Freeman WJ. (2003) A neurobiological theory of meaning in perception. Part 1. Information and meaning in nonconvergent and nonlocal brain dynamics. *Int. J. Bifurc. Chaos* 13: in press.

Freeman WJ, Schneider W (1982) Changes in spatial patterns of rabbit olfactory EEG with conditioning to odors. *Psychophysiol.* 19: 44-56.

Freeman WJ, Barrie JM (2000) Analysis of spatial patterns of phase in neocortical gamma EEGs in rabbit. *J. Neurophysiol.* 84: 1266-1278.

Freeman WJ, Gaál G and Jörsten R. (2003) A neurobiological theory of meaning in perception. Part 2. Multiple cortical areas synchronize without loss of local autonomy. *Int. J. Bifurc. Chaos* 13: in press.

Gibson JJ (1979) *The Ecological Approach to Visual Perception.* Boston: Houghton Mifflin.

Gloor P (1997) *The Temporal Lobe and the Limbic System.* New York: Oxford University Press.

Goltz FL (1892) Der Hund ohne Grosshirn. Siebente Abhandlung über die Verrichtungen des Grosshirns. *Pflügers Archiv* 51: 570-614.

Gray CM (1994) Synchronous oscillations in neuronal systems: mechanisms and functions. *J. Comp. Neurosci.* 1: 11-38.

Grossberg S (1982) *Studies of Mind and Brain: Neural Principles of Learning, Perception Development, Cognition and Motor Control.* Boston: D. Reidel.

Haken H (1983) *Synergetics: An Introduction.* Berlin: Springer.

Hardcastle VG (1994) Psychology's binding problem and possible neurobiological solutions. *J. Consciousness Studies* 1: 66-90.

Hebb DO (1949) *The Organization of Behavior.* New York: Wiley.

Helmholtz HLF von (1872) *Handbuch der physiologischen Optik. Vol. III.* Leipzig: L. Voss (1909).

Herrnstein RJ & Murray C (1994) *The Bell Curve.* New York: The Free Press.

Herrick CJ (1948) *The Brain of the Tiger Salamander.* Chicago IL: University of Chicago Press.

Hull CL (1943) *Principles of Behavior, An Introduction to Behavior Theory.* New York: Appleton-Century

Hume D (1739) *Treatise on Human Nature.* London: J Noon.

Jackson JH (1931) *Selected writings of John Hughlings Jackson.* Taylor J, Walshe FMR, Holmes G (eds.). London: Hodder & Stoughton.

James W (1879) Are we automata? *Mind* 4: 1-21.

James W (1893) *The Principles of Psychology.* New York. H. Holt.

James, W (1890) The Principles of Psychology p. 1264 in: *The Works of William James*, 2, p. 1264, Burkhardt FH (ed. 19 vols). Cambridge MA: Harvard University Press.

Kay LM, Freeman WJ (1998) Bidirectional processing in the olfactory-limbic axis during olfactory behavior. *Behav. Neurosci.* 112: 541-553.

König P, Schillen TB (1991) Stimulus-dependent assembly formation of oscillatory responses: I. synchronization. *Neural Comp.* 3, 155-166.

Kozma R, Freeman WJ (2001) Chaotic resonance: Methods and applications for robust classification of noisy and variable patterns. *Int. J. Bifurc. Chaos* 10: 2307-2322.

Lehmann D, Michel CM (1990) Intracerebral dipole source localization for FFT power maps. *Electroenceph. Clin. Neurophysiol.* 76: 271-276.

Lesse H (1957) Amygdaloid electrical activity during a conditioned response. *Proc. 4th Intern. Congress EEG clin. Neurophysiol.*, Brussels, 99-100.

Libet B (1994) *Neurophysiology of Consciousness: Selected Papers and New Essays.* Boston MA: Birkhauser.

Lorente de Nó R (1934) Studies on the structure of the cerebral cortex. I The area entorhinalis. *J. für Psychologie und Neurologie* 45: 381-438.

McCulloch WS (1967) *Embodiments of Mind.* Cambridge MA: MIT Press.

McNaughton BL (1993) The mechanism of expression of long-term enhancement of hippocampal synapses: Current issues and theoretical implications. *Ann. Rev. Physiol.* 55: 375-96.

Menand L (2001) *The Metaphysical Club*. New York: Farrar, Straus & Giroux, 2001.

Merleau-Ponty M (1942) *The Structure of Behavior* (AL Fischer, Trans.). Boston: Beacon Press (1963).

Mill JS (1843) *Of Liberty and Necessity, Ch. II, Book VI. A System of Logic.* London UK: Longmans, Green 18th ed. (1965)

Mill JS (1873) *Autobiography*. New York: Columbia University Press (1924).

Miltner WHR, Barun C, Arnold M, Witte H, Taub E (1999) Coherence of gamma-band EEG activity as a basis for associative learning. *Nature* 397: 434-436.

Müller MM (2000) Hochfrequente oszillatorische Aktivitäten im menschlichen Gehirn. *Zeitschrift für Exper. Psychol.* 47: 231-252.

Ohl, F.W., Scheich, H., Freeman W.J. (2001) Change in pattern of ongoing cortical activity with auditory category learning. *Nature* 412: 733-736.

O'Keefe J & Nadel L (1978) *The Hippocampus as a Cognitive Map*. Oxford UK: Clarendon.

Panksepp, J. (1998) *Affective Neuroscience.* Oxford University Press.

Penrose R (1994) *Shadows of the Mind.* Oxford UK: Oxford University Press.

Piaget J (1930) *The child's conception of physical causality*. New York: Harcourt, Brace.

Popper KR (1990) *A World of Propensities*. Bristol UK: Thoemmes.

Prigogine I (1980) *From Being to Becoming: Time and Complexity in the Physical Sciences*. San Francisco: Freeman.

Principe, J.C., Tavares, V.G., Harris, J.G. and Freeman W.J. (2001) Design and implementation of a biologically realistic olfactory cortex in analog VLSI. *Proceedings IEEE* 89: 1030-1051.

Putnam H (1990) *Realism With a Human Face*. Cambridge MA: Harvard University Press.

Rodriguez E, George N Lachaux J-P, Martinerie J, Renault B, Varela F (1999) Perception's shadow: long-distance synchronization of human brain activity. *Nature* 397: 430-433.

Roelfsema, P.R., Engel, A.K., König, P. & Singer, W. (1997) Visuomotor integration is associated with zero time-lag synchronization among cortical areas. *Nature* 385, 157-161.

Rolls ET, Miyashita Y, Cahusac PBM, Kesner RP, Niki H, Feigenbaum JD & Bach L (1989) Hippocampal neurons in the monkey with activity related to the place in which the stimulus is shown. *J. Neurosci.* 9: 1835-1845.

Roth G (1987) *Visual Behavior in Salamanders*. Berlin: Springer-Verlag

Searle JR (1992) *The Rediscovery of Mind.* Cambridge MA: MIT Press.

Sherrington CS (1940) *Man on his Nature.* Oxford UK: Oxford University Press.

Singer W, Gray CM (1995) Visual feature integration and the temporal correlation hypothesis. *Ann. Rev. Neurosci.* 18: 555-586.

Skarda CA, Freeman WJ (1987) How brains make chaos in order to make sense of the world. *Behav. Brain Sci.* 10: 161-195.

Skinner BF (1969) *Contingencies of Reinforcement; A Theoretical Analysis*. New York: Appleton-Century-Crofts

Smart A, German P, Oshtory S, Gaál G, Barrie JM, Freeman WJ (1997) Spatio-temporal analysis of multi-electrode cortical EEG of awake rabbit. *Soc. Neurosci. Abstracts* 189.13.

Sperry RW (1950) Neural basis of the spontaneous optokinetic response. *J. Comp. Physiology* 43: 482-489.

Tallon-Baudry C, Bertrand O, Peronnet F, Pernier J (1998). Induced gamma-band activity during the delay of a visual short-term memory task in humans. *J. Neurosci.* 18: 4244-4254.

Thelen E, Smith LB (1994) *A Dynamic Systems Approach to the Development of Cognition and Action.* Cambridge MA: MIT Press.

Titchener EB (1907) *An Outline of Psychology.* New York: Macmillan.

Traub RD, Whittington MA, Colling SB, Buzsaki G, Jefferys JGR (1996) A mechanism for generation of long-range synchronous fast oscillations in the cortex. *Nature* 383: 621-624.

Tsuda I. (2001) Toward an interpretation of dynamics neural activity in terms of chaotic dynamical systems. *Behav. & Brain Sci.* 24: 793-847.

von Holst E & Mittelstädt H (1950) Das Reafferenzprinzip (Wechselwirkung zwischen Zentralnervensystem und Peripherie). *Naturwissenschaften* 37: 464-476.

von Neumann J (1958) *The Computer and the Brain.* New Haven CT: Yale University Press.

Whitehead AN (1938) *Modes of Thought*. New York: Macmillan.

Wilson MA & McNaughton BL (1993) Dynamics of the hippocampal ensemble code for space. *Science* 261: 1055-1058.

In: New Research on Chaos and Complexity ISBN 1-60021-275-1
Editors: F. Orsucci and N. Sala, pp. 47-64

Chapter 4

THE STRUCTURAL EQUATIONS TECHNIQUE FOR TESTING HYPOTHESES IN NONLINEAR DYNAMICS: CATASTROPHES, CHAOS, AND RELATED DYNAMICS

Stephen J. Guastello [*]
Marquette University, Milwaukee, WI 53201-1881 USA,

ABSTRACT

This article summarizes the central analysis points for testing hypotheses concerning catastrophe models, chaos, and related dynamics as they are encountered in the social sciences. Two model classes are considered: the catastrophe models for discontinuous change processes, and the exponential series for continuous change. Some frequently asked questions concerning the types of and amount of data that are required for viable analyses are parenthetically answered here.

One of the gripping problems in nonlinear dynamics is difficulty in testing hypotheses with real data, especially the type of data that are collected in the

* *Send correspondence concerning the manuscript to:* Stephen J. Guastello, Ph.D., Dept. Psychology, Marquette University, P.O. Box 1881, Milwaukee, WI 53201-1881 USA, Tel: 414-288-6900, stephen.guastello@marquette.edu

social sciences. In spite of controversies and tales of woe that persist in discussions on listservers, reliable techniques exist, and indeed some have been in use for more than twenty years. The techniques described in this article are predicated on the assumption that the researcher has a viable hypothesis concerning one or more specific models that require explicit testing. A basic understanding of attractors, bifurcations, chaos, catastrophes, and self-organization is necessarily assumed here.

It is a good idea, nonetheless, to review why anyone would bother studying nonlinear dynamic systems (NDS) at all. There are four basic reasons: (a) NDS theory provides concepts that explain changes that occur over time. (b) NDS theory allows for structural comparisons of models across situations that may be very different in their outward appearances. (c) NDS solutions to problems, when they have been adopted, provide better explanations of data if and when they are expressed by R^2, or percentage of variance accounted for by a nonlinear model (Guastello, 1992a, 1995, 2002). (d) NDS provides effective solutions to theoretical questions; see Guastello (2001a) for a survey of progress on topics in psychology.

OVERVIEW

The structural equations technique begins be defining a model in the form of an equation, then testing it statistically with real (as opposed to simulated) data. The analysis separates the deterministic portion of the data from noise. "Noise" here denotes that portion of the data variance that is not explained by the deterministic equation. Social scientists will recognize this model-versus-noise approach as "business as usual." This technique contrasts, however, with a prevailing habit in the physical sciences, which works in the opposite fashion: Separate the noise first, then make calculations on what remains (e.g. Kanz & Schreiber, 1997).

Two series of hierarchical models are considered here. The first is the catastrophe models for discontinuous change. The catastrophes, which were originally introduced by Thom (1975) have received renewed attention because of their relevance to self-organized systems (Kelso, 1995; Zhang, 2002). The second set involved exponential models for continuous change and includes a test for the Lyapunov exponent which distinguishes between chaos and non-chaotic dynamics. The latter set was introduced by Guastello (1995) and built on previous work by May and Oster (1976), Wiggins (1988) and numerous other contributors to the field of NDS.

Each model in a hierarchy subsumes properties of the simpler models. Each progressively complex model adds a new dynamical feature. This article covers models involving only one order parameter (dependent measure). Two-parameter models can be tested as well, but the reader is directed to Guastello (1995) to see how those extrapolations can be accomplished.

The following sections of this article address the type of data and amounts that are required; probability functions, location, and scale; the structure of behavioral measurements; the catastrophe model group, which can be tested through power polynomial regression; the exponential series of models, which can be tested through nonlinear regression; and catastrophe models that are testable as static probability functions through nonlinear regression.

Types and Amount of Data

The procedures that follow require dependent measures (order parameters) that are measured at two points in time at least. One may have many entities that are measured at two points in time, or one long time series of observations from one entity. Alternatively, one may have an ensemble of shorter time series taken from several entities.

In general it is better to have a smaller number of observations that cover the full dynamical character of a phenomenon than to have a large number of observations that cover the underlying topology poorly. Because these are statistical procedures, all the usual rules and caveats pertaining to statistical power apply. The simplest models can be tested with 50 data points, and sometimes fewer of them, if (a) a good model of the phenomenon in question, (b) reliable measurements, (c) only one or two regression parameters to estimate. In all cases, more data is better than less data so long as the data are actually covering all the nonlinear dynamics that are thought to exist in the system.

Statistical Power

The calculation of statistical power for ordinary multiple regression depends on the intended effect size, overall sample R^2, population R^2, the number of independent variables, the degree of correlation among the independent variables, and occasionally the assumption that all independent variables are equally weighted. Not surprisingly there are different rubrics for determining proper sample size.

According to the Cohen (1988), the appropriate sample size to detect a "medium" effect size of .15 for one of the independent variables with a power of .80 is 52 plus the number of estimated parameters (Maxwell, 2000). Thus the required sample size would be 58 observations for a six-parameter model. According to a similar rubric by Green (1991), a sample size of 110 should to detect an effect size of .075 for one independent variable with a power of .80. Neither rubric takes into account that the odds of finding a smaller partial correlation increase to the extent that the overall R^2 is large. According to Maxwell (2000), the odds of detecting one of the effects within a multiple regression model drops sharply as the correlations among independent variables increases.

One should bear in mind, however, that the calculation of statistical power for nonlinear regression is still generally uncharted territory. It is thus necessary to rely on the rubrics for linear models. If there is sufficient statistical power for the linear comparison models which, in the past, have been generally weaker in overall effect size than nonlinear models when the nonlinear model was held true, there should not be much concern with the statistical power of the nonlinear models. On the other hand, the power for specific effects within a nonlinear model probably depends on whether the regression parameter is associated with an additive, exponential, multiplicative, or other type of mathematical operator.

Optimal Time Lag

Put simply, the time lag between observations is optimal if it reflects the real time frame in which data points are generated. For instance, catastrophe models are usually lagged "before" and "after" a discrete event. Macroeconomic variables might be studied best at lags equal to an economic quarter of the year (e.g. Guastello, 1999a).

Probability Density Functions

It is convenient that any differential function can be transformed into a probability density function pdf using the Wright-Ito transformation. The variable y in Eq. 1. is a dependent measure that exhibits the dynamical character under study; y is then transformed into z with respect to *location* (λ) and *scale* (σ_s, Eq. 2).

$$\text{pdf}(z) = \xi \exp[\, - \textstyle\int f(z)]; \tag{1}$$

$$z = (\,y - \lambda\,) \,/\, \sigma_s\,. \tag{2}$$

Location

In most discussions of probability functions, "location" refers to the mean of the function. In dynamics the pdf is a member of an exponential family of distributions and is asymmetrical, unlike the so-called normal distribution. Thus the location parameter for Eq. 2 is the lower limit of the distribution, which is the lowest observed value in the series. The transformation in Eq. 2 has the added advantage of fixing a zero point and thus transforming measurements with interval scales (common in the social sciences) to ratio scales. A fixed location point defines where the nonlinear function is going to start.

Scale

The scale parameter in common discussions of pdfs is the standard deviation of the distribution. The standard deviation is used here also. The use of the scale parameter later on while testing structural equations serves the purpose of eliminating bias between two or more variables that are multiplied together. Although the results of linear regression are not affected by values of location and scale, nonlinear models are clearly affected by the transformation.

Occasionally one may obtain a better fit using the alternative definition of scale in Eq. 3, which measures variability around statistical modes rather than around a mean:

$$\sigma_s = \left[\sqrt{\Sigma \Sigma \left(y_{m-}^{m+} - m \right)^2} \right] / N - M \quad \text{y=1 m=1} \tag{3}$$

To use it, the distribution must be broken into sections, each section containing a mode or an antimode. The values of the variable around a mode will range from *m*- to *m*+ as depicted in Eq. 3.

Corrections for location and scale should be made on control variables as well as for dependent measures. The ordinary standard deviation is a suitable measure of scale for control variables.

Structure of Behavioral Measurements

In the classic definition, a measurement consists of true scores (T) plus error (e). The variance structure for a population of scores is thus:

$$\sigma^2(X) = \sigma^2(T) + \sigma^2(e). \quad (4)$$

The classical assumption is that all errors are independent of true scores and all other errors.

In nonlinear dynamics our true score is the result of a linear (L) and nonlinear deterministic process (NL), dependent error (DE), and independent error (IIDE):

$$\sigma^2(X) = \sigma^2(L) + \sigma^2(NL\text{-}L) + \sigma^2(DE) + \sigma^2(IIDE) \quad (5)$$

"Independent error" in conventional psychometrics is known as independently and identically distributed (IID) error in the NDS literature. Importantly the non-IID error is a result of the nonlinear deterministic process (Brock, Hseih, & LeBaron, 1991).

Catastrophe Models

The analytic method that follows requires the polynomial form of multiple linear regression. The analysis can be performed with most any standard statistical software package. Several concepts for hypothesis testing carry through to subsequent analyses of other dynamics.

The set of catastrophe models was the result of the classification theorem by Thom (1975): Given certain constraints, all discontinuous changes in events can be described by one of seven elementary models. Four of the models contain one order parameter; this is the cuspoid series: fold model which has one control parameter, cusp model with two control parameters, swallowtail model with three control parameters, butterfly model with four control parameters. The remaining three models, known as the umbilic group, contain two order parameters. The instructions that follow pertain to the cuspoid group.

The process of hypothesis testing begins by choosing a model that appears to be closest to the phenomenon under investigation. Because the cusp is the most often used model, the following remarks are framed in terms of the cusp model. The cusp (Fig. 1) depicts two stable states of behavior and requires two control variables. The two stable states are separated by a bifurcation manifold or separatrix. The asymmetry parameter governs how close the system is to a sudden discontinuous change of events. The bifurcation parameter governs how large the change will be. Two or more experimental variables may be hypothesized for each control parameter without changing the basic model or analytic procedure.

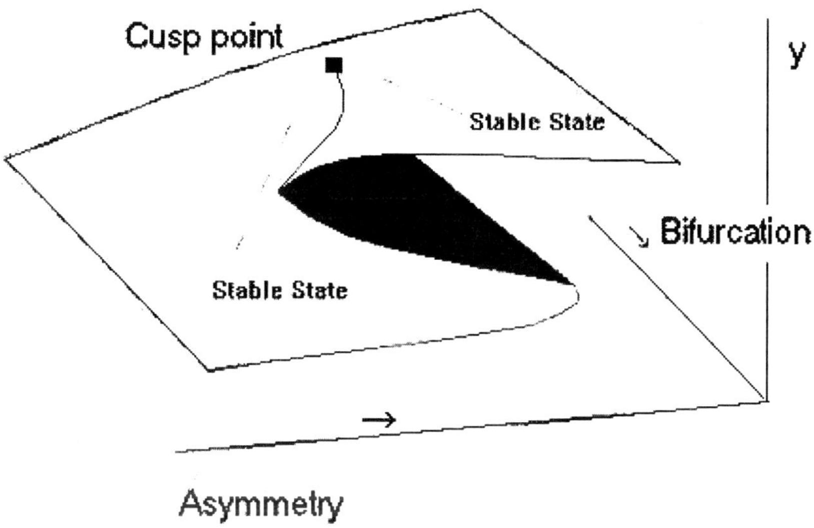

Fig. 1. Response surface for the cusp catastrophe model.

Nonlinear Statistical Model

The deterministic equation for the cusp is shown in Eq. 6, and followed by its probability density function using the Wright-Ito transformation in Eq. 7:

$$\delta f(y)/\delta y = y^3 - by - a, \tag{6}$$

$$\text{pdf}(z) = \xi\, e^{[-z4/4 + bz2/2 + az]}. \tag{7}$$

Figure 2 shows an example of what a cusp pdf could look like using real data (Guastello, 2002, p. 136).

Next, we take the deterministic equation for the cusp response surface, insert regression weights, and a quadratic term:

$$\Delta z = \beta_0 + \beta_1 z_1^3 + \beta_2 z_1^2 + \beta_3 b z_1 + \beta_4 a, \quad (8)$$

The quadratic term is an additional correction for location. The dependent measure Δz denotes a change in behavior over two subsequent points in time.

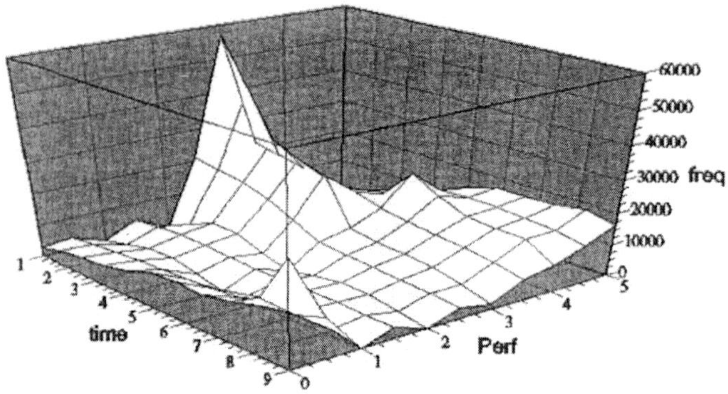

Fig. 2. Example of a cusp pdf for a personnel selection and turnover problem (Guastello, 2002). Reprinted with permission from Lawrence Erlbaum Associates.

Several hypotheses are being tested in the power polynomial equation (Eq. 8). There is the F test for the model overall; retain the R^2 coefficient and save it for later use. There are t tests on the beta weights; they denote which parts of the model account for unique portions of variance.

Some model elements are more important than other elements. The cubic term expresses whether the model is consistent with cusp structure; the correct level of complexity for a catastrophe model is captured by the leading power term. If there is a cusp structure, then one must identify a bifurcation variable as represented by the $\beta b z_1$ term. A cusp hypothesis is not complete without a bifurcation term; shabby results may be expected otherwise. The asymmetry term βa is important in the model, but failing to find one does not negate the cusp structure if the cubic and bifurcation elements are present. The lack of an asymmetry term only means that the model is not complete.

The quadratic term is the most expendable. It is not part of the formal deterministic cusp structure. Rather it is an additional correction for location (Cobb, 1981a). In the event that unique weights are not obtained for all model components, delete the quadratic term and test the remaining elements again.

Note the procedural contrast with linear regression analysis: In common linear regression, when a variable does not attain a significant weight, we simply delete that variable. In NDS, we delete variables based on their relative importance to the structural model. In linear analyses, there is only a linear structure under consideration, so particular variables are then kept or discarded. In nonlinear analyses, different variables may be playing different structural roles.

Linear Comparison Models

Next construct Eqs. 9 and 10 and compare their R^2 coefficients against the R^2 that was obtained for the cusp:

$$y_2 = B_0 + B_1 y_1 + B_2 a + B_3 b, \tag{9}$$

$$\Delta y = B_0 + B_1 a + B_2 b. \tag{10}$$

Next evaluate the elements of the cusp model. If all the necessary parts of the cusp are significant, and the R^2 coefficients compare favorably, then a clear case of the cusp has been obtained.

Exponential Model Series

This section describes a series of models that exhibit continuous but nevertheless interesting change. The model structures are functions of the Naperian constant e. They produce, among other things, the Lyapunov exponent, which is a test for chaos and a value comparable to the fractal dimension.

Nonlinear regression is required the test this series of models. Nonlinear regression may be familiar to biologists, but is probably much less familiar to social scientists at the present time. The hierarchical series of models ranges from simplest to complex as follows: (a) simple Lyapunov exponent, (b) Lyapunov with additional fitting constants, (c) May-Oster model with the bifurcation parameter unknown, (d) model with an explicitly hypothesized bifurcation model, and (e) models with two or more order parameters.

Lyapunov Models

The simplest model predicts behavior z_2 from a function of z_1. Note that the corrections for location and scale apply here as well:

$$z_2 = e^{(\theta_1 z_1)} \qquad (11)$$

The nonlinear regression weight θ_1 is located in the exponent. θ_1 is also the Lyapunov exponent. It is a measure of turbulence in the time series. If θ_1 is positive, then chaos is occurring. If θ_1 is negative, then a fixed point or periodic dynamic is occurring. D_L is an approximation of the fractal dimension (Ott, Sauer, & Yorke, 1994):

$$D_L = e^{\theta_1}. \qquad (12)$$

The second model in the series is the same as the first except that two constants have been introduced to absorb unaccounted variance. The Lyapunov exponent is now designated as θ_2:

$$z_2 = \theta_1 e^{(\theta_2 z_1)} + \theta_3 . \qquad (13)$$

In nonlinear regression it is necessary to specify the placement of constants in a model. Unlike the general linear model, constants in nonlinear models can appear anywhere at all. Hence θ_1 and θ_3 are introduced in Eq. 13.

The suggested strategy here is to start with the second model (Eq. 13). If statistical significance is not obtained for all three weights, delete θ_1 and try again. If that result is not good enough, drop the additive constant θ_3 and return to the simplest model of the series (Eq. 11).

Bifurcation Models

The third level of model is shown in Eq. 14. Note the introduction of z_1 between θ_1 and e:

$$z_2 = \theta_1 z_1 e^{\theta_2 z_1} + \theta_3. \qquad (14)$$

Eq. 14 tests for the presence of a variable that is possibly changing the dynamics of a model. For instance, some learning curves could be sharper than others. A positive test for the model indicates that a variable is present, but its identity is not yet known.

The computation of dimension is similar to that of previous exponential models, except that a value of 1 must be added to account for the presence of a bifurcation variable:

$$D_L = e^{\theta_2} + 1. \tag{15}$$

At the fourth level of complexity, the researcher has a specific hypothesis for the bifurcation variable, which is designated as *c* in Eq. 16:

$$z_2 = \theta_1 c\, z_1\, e^{\theta_2 z_1} + \theta_3. \tag{16}$$

Linear Contrasts

As in the case of the catastrophes, we test the R^2 for the nonlinear regression model against that of the linear alternatives, such as

$$y_2 = B_0 + B_1 y_1 \tag{17}$$

or

$$y_2 = B_0 + B_1 t, \tag{18}$$

where *t* is time.

Tips for Using Nonlinear Regression

On the model parameter command line (or comparable command in statistical packages), the names of the regression weights are specified with initial values. Use either the initial values of 0.5 or pick your own. When in doubt, give the initial weights equal value. Often it does not matter whether the iterative computational procedure states off with equal weights or not. If the model results are not affected by the initial values, then the resulting model is more robust than what would be the case otherwise.

If there is an option to choose constrained versus unconstrained nonlinear regression, use the unconstrained option, which is typically the default. Constraints indicate that the researcher expects the values of parameters to remain within numerical boundaries that have been pre-determined. Occasionally there may be a good rationale for constraining parameters, but they would be specific to the problem when they exist.

If there is an option to choose least squares or maximum likelihood error term specification be forewarned: Maximum likelihood is more likely to capitalize on chance aspects of the pdf, and is thus more likely to return a significant result. I use least squares.

If the results of a nonlinear regression analysis are so poor that they produce a negative R^2, do not be alarmed. Just treat the negative R^2 as if it were .00.

When testing for significance, the tests on the weights are very important. Some researchers value them more than the overall R^2. Tests for weights are made using the principle of confidence intervals. An alpha level of $p < .05$ is regarded as unilaterally sufficient. A nonsignificant weight with a high overall R^2 could be the result of a high correlation among the parameter estimates; this condition is akin to multicollinearity in ordinary linear models.

Testing Catastrophes Through Nonlinear Regression

The two strategies previously delineated can now be combined for some special circumstances. Sometimes one might obtain a pdf that bears a strong resemblance to that of an elementary catastrophe, and it is logical to frame a hypothesis as to whether that association is true or false. In another situation, there may be a catastrophe process occurring, but all the time-1 measurements are the same value of 0.00.

In both types of situations it would be good to test a hypothesis concerning the catastrophe distribution. This type of test is described below for a situation that involves that swallowtail catastrophe model. The swallowtail response surface is shown in Fig 3. Because the response surface is four-dimensional, it must be presented in two 3-dimensional sections.

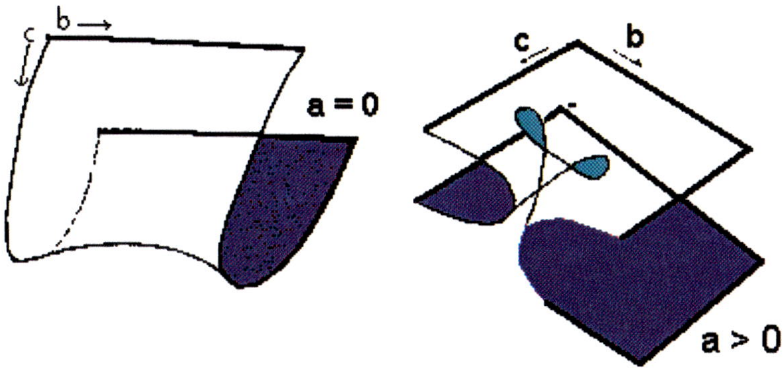

Fig 3. Swallowtail catastrophe response surface (Guastello, 2002). Reprinted with permission from Lawrence Erlbaum Associates.

The equation for the response surface is shown in Eq. 19:

$$\delta f(y)/\delta y = y^4 - cy^2 - by - a, \tag{19}$$

The swallowtail pdf is shown in Eq. 20 and Fig. 4:

$$\text{pdf}(z) = \xi \exp[-z^5/5 + z^4/4 + cz^3/3 + bz^2/2 + az]. \tag{20}$$

The pdf is tested as a nonlinear regression model in Eq. 21:

$$\text{pdf}(z) = \xi \exp[-\theta_1 z^5 + \theta_2 z^4 - \theta_3 cz^3 - \theta_4 bz^2 - \theta_5 az]; \tag{21}$$

Note where the regression weights are inserted in Eq. 21. θ_i is also treated as a regression weight. Pdf(z) is the cumulative probability of *z* within the distribution.

If the control variables are not known yet (it is necessary to have hypotheses about them in the polynomial regression models), variables *a*, *b*, and *c*, in Eq. 21 can be ignored. One would thus be treating them essentially as constants (or part of the regression weight).

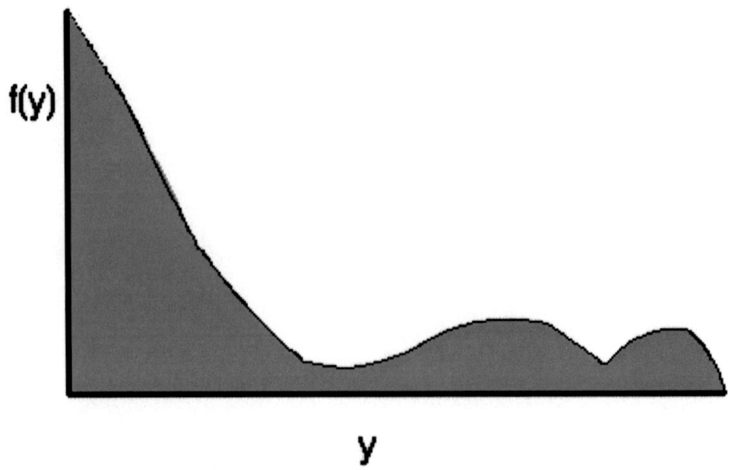

Fig. 4. Slice of the swallowtail pdf.

An actual (slice of a) swallowtail PDF appears in Fig. 5. The jagged contour is real data. The smooth contour is based on estimated values that were obtained from an analysis using Eq. 22 (Guastello, 1998a):

$$\text{freq}(y) = \beta_0 + \beta_1 y + \beta_2 y^2 + \beta_3 y^3 + \beta_4 y^4 \tag{22}$$

Eq. 22 is a polynomial regression model where the *frequency of y* is a function of the value of *y*. This is the easiest way to make the smooth contour and identify critical points that correspond to statistical modes and antimodes. The jagged contour in Fig. 5 is the actual pdf of leadership ratings from an emergent leadership experiment. The modes are located at values of 0, 9, and 12, with antimodes at 7 and 11. The estimated values in the smooth contour denote modes at values of 0, 9, and 16, with antimodes at 5, and 13.

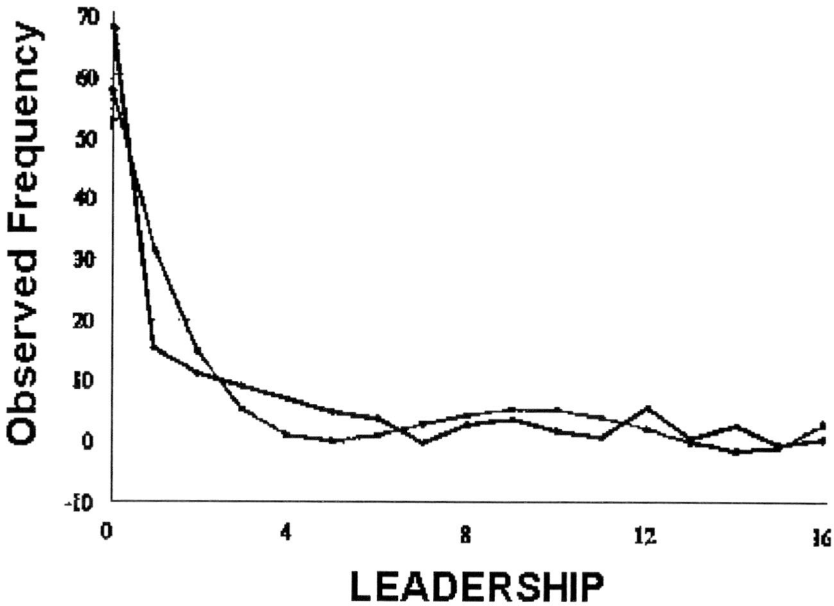

Fig. 5. Comparison of actual and predicted pdf data from a swallowtail catastrophe problem (Guastello, 1998a). Reprinted with permission from Kluwer Academic Publishers.

EXAMPLES ON RECORD

Examples of analyses using the polynomial regression method for catastrophes date back to Guastello (1982). Recent examples, however, can be found in Guastello (1995, 2002), Guastello, Gershon, & Murphy (1999), Clair, (1998); Lange (1999), and Byrne, Mazanov, and Gregson (2001).

Examples of analyses using the nonlinear regression method for chaos and related exponential models date back to Guastello (1992b). More recent examples can be found in Guastello (1995, 1998b, 1999a, 1999b, 2001b, 2002), Guastello and Philippe (1997), Guastello and Guastello (1998). Guastello and Johnson (1999), Guastello, Johnson, and Rieke (1999), Guastello and Bock (2001), Rosser, Rosser, Guastello, and Bond (2001), and Guastello and Bond (2004). For examples that compared dimensionality estimates made through nonlinear regression with values obtained by other means, see Johnson and Dooley (1996) and Guastello and Philippe (1997).

Examples of analyses using the nonlinear regression method for testing catastrophe pdfs are sparse, although the method was proposed by Cobb (1981a, 1981b). More recent examples can be found in Hanges, Braverman, & Rentch (1991), Guastello (1998a, 2002), and Zaror and Guastello (2000).

REFERENCES

Brock, W. A., Hseih, D. A., LeBaron, B. (1991). *Nonlinear dynamics, chaos, and instability: Statistical theory and economic evidence*. Cambridge, MA: MIT Press.

Byrne, D. G., Mazanov, J., & Gregson, R. A. M. (2001). A cusp catastrophe analysis of changes to adolescent smoking behavior in response to smoking prevention programs. *Nonlinear Dynamics, Psychology, and Life Sciences, 5*, 115-138.

Clair, S. (1998). A cusp catastrophe model for adolescent alcohol use: An empirical test. *Nonlinear Dynamics, Psychology, and Life Sciences, 2*, 217-241.

Cobb, L.(1981a). Multimodal exponential families of statistical catastrophe theory. In C. Taillie, G. P. Patel, & B. Baldessari (Eds.), *Statistical distributions in scientific work* (vol. 6, pp. 67-90). Hingam, MA: Reidel.

Cobb, L. (1981b). Parameter estimation for the cusp catastrophe model. *Behavioral Science, 26*, 75-78.

Cohen, J. (1988). *Statistical power analysis for the behavioral sciences*. Hillsdale, NJ: Lawrence Erlbaum Associates.

Green, B. F. (1991). How many subjects does it take to do a regression analysis? *Multivariate Behavioral Research, 12*, 263-288.

Guastello, S. J. (1982a). Color matching and shift work: An industrial application of the cusp-difference equation. *Behavioral Science, 27*, 131-137.

Guastello, S. J. (1992a). Clash of the paradigms: A critique of an examination of the polynomial regression technique for evaluating catastrophe theory hypotheses. *Psychological Bulletin, 111*, 375-379.

Guastello, S. J. (1992b). Population dynamics and workforce productivity. In M. Michaels (Ed.), *Proceedings of the annual conference of the Chaos Network: The second iteration* (pp. 120-127). Urbana, IL: People Technologies.

Guastello, S. J. (1995). *Chaos, catastrophe, and human affairs: Applications of nonlinear dynamics to work, organizations, and social evolution*. Mahwah, NJ: Lawrence Erlbaum Associates.

Guastello, S. J. (1998a). Self-organization and leadership emergence. *Nonlinear Dynamics, Psychology, and Life Sciences, 2*, 303-316.

Guastello, S. J. (1998b). Creative problem solving groups at the edge of chaos. *Journal of Creative Behavior, 32*, 38-57.

Guastello, S.J. (1999a). Hysteresis, bifurcation, and the natural rate of unemployment. In E. Elliott & L. D. Kiel, (Eds.). *Nonlinear dynamics, complexity and public policy* (pp. 31-46). Commack, NY: Nova Science.

Guastello, S. J. (1999b). Hierarchical dynamics affecting work performance in organizations. In W. Tschacher & J-P. Dauwaulder (Eds.), *Dynamics, synergetics and autonomous agents* (pp. 277-302). Singapore: World Scientific.

Guastello, S. J. (2001a). Nonlinear dynamics in psychology. *Discrete Dynamics in Nature and Society, 6*, 11-29.

Guastello, S. J. (2001b). Attractor stability in unemployment and inflation rates. In Y. Aruka (Ed.) *Evolutionary controversies in economics: A new transdiscipinary approach* (pp. 89-99). Tokyo: Springer-Verlag.

Guastello, S. J. (2002). *Managing emergent phenomena: Nonlinear dynamics in work organizations*. Mahwah, NJ: Lawrence Erlbaum Associates.

Guastello, S. J. & Bond R. W. (2004). The Stag Hunt coordination game and emergency response teams. *Nonlinear Dynamics, Psychology, and Life Sciences, 8*, 345-374.

Guastello, S. J., & Bock, B. R. (2001). Attractor reconstruction with principal components analysis: Application to work flows in hierarchical organizations. *Nonlinear Dynamics, Psychology, and Life Sciences, 5, 175-192.*

Guastello, S. J., Gershon, R. M., & Murphy, L. R. (1999). Catastrophe model for the exposure to blood-borne pathogens and other accidents in health care settings. *Accident Analysis and Prevention, 31, 739-750.*

Guastello, S. J., & Guastello, D. D. (1998). Origins of coordination and team effectiveness: A perspective from game theory and nonlinear dynamics. *Journal of Applied Psychology, 83*, 423-437.

Guastello S. J., & Johnson, E. A. (1999). The effect of downsizing on hierarchical work flow dynamics in organizations. *Nonlinear Dynamics, Psychology, and Life Sciences, 3*, 347-378.

Guastello, S. J., Johnson, E. A., & Rieke, M. L. (1999). Nonlinear dynamics of motivational flow. *Nonlinear Dynamics, Psychology, and Life Sciences, 3*, 259-273.

Guastello, S. J., & Philippe, P. (1997). Dynamics in the development of large information exchange groups and virtual communities. *Nonlinear Dynamics, Psychology, and Life Sciences, 1*, 123-149.

Hanges, P. J., Braverman, E. P., Rentch, J. R. (1991). Changes in raters' perception of subordinates: A catastrophe model. *Journal of Applied Psychology, 76*, 878-888.

Johnson, T. L., & Dooley, K. J. (1996). Looking for chaos in time series data. In W. Sulis and A. Combs (Eds.), *Nonlinear dynamics in human behavior* (pp. 44-76). Singapore: World Scientific.

Kanz, H., & Schreiber, T. (1997). *Nonlinear time series analysis*. New York: Cambridge University Press.

Kelso, J. A. S. (1995*). Dynamic patterns: Self-organization of brain and behavior*. Cambridge, MA: MIT Press.

Lange, R. (1999). A cusp catastrophe approach to the prediction of temporal patterns in the kill dates of individual serial murderers. *Nonlinear Dynamics, Psychology, and Life Sciences, 3*, 143-159.

Maxwell, S. E. (2000). Sample size and multiple regression analysis. *Psychological Methods, 5*, 343-458.

May, R. M., & Oster, G. F. (1976). Bifurcations and dynamic complexity in simple ecological models. *American Naturalist, 110*, 573-599.

Ott, E., Sauer, T., & Yorke, J. A. (Eds.). (1994). *Coping with chaos*. New York: Wiley.

Rosser, J. B., Rosser, M. V., Guastello, S. J., & Bond, R. W. Jr. (2001). Chaotic hysteresis and systemic economic transformation: Soviet investment patterns. *Nonlinear Dynamics, Psychology, and Life Sciences, 5*, 345-368.

Thom, R. (1975). *Structural stability and morphegenesis*. New York: Benjamin-Addison- Wesley.

Wiggins, S. (1988). *Global bifurcations and chaos*. New York: Springer-Verlag.

Zaror, G., & Guastello, S. J. (2000). Self-organization and leadership emergence: A cross- cultural replication. *Nonlinear Dynamics, Psychology, and Life Sciences, 4, 113-119.*

Zhang, W-B. (2002). Theory of complex systems and economic development. *Nonlinear Dynamics, Psychology, and Life Sciences, 6*, 83-102.

In: New Research on Chaos and Complexity ISBN 1-60021-275-1
Editors: F. Orsucci and N. Sala, pp. 65-96

Chapter 5

SYNCHRONIZATION OF OSCILLATORS IN COMPLEX NETWORKS

Louis M. Pecora[1] and Mauricio Barahona[2]
[1]Code 6343
Naval Research Laboratory
Washington, DC 20375, USA
[2]Department of Bioengineering, Mech. Eng. Bldg., Imperial College of STM,
Exhibition Road, London SW7 2BX, UK

ABSTRACT

We introduce the theory of identical or complete synchronization of identical oscillators in arbitrary networks. In addition, we introduce several graph theory concepts and results that augment the synchronization theory and tie is closely to random, semirandom, and regular networks. We then use the combined theories to explore and compare three types of semirandom networks for their efficacy in synchronizing oscillators. We show that the simplest *k*-cycle augmented by a few random edges or links appears to be the most efficient network that will guarantee good synchronization.

I. Introduction

In the past several years interest in networks and their statistics has grown greatly in the applied mathematics, physics, biology, and sociology areas. Although networks have been structures of interest in these areas for some time recent developments in the construction of what might be called structured or semirandom networks has provoked increased interest in both studying networks and their various statistics and using them as more realistic models for physical or biological systems. At the same time developments have progressed to the point that the networks can be treated not just as abstract entities with the vertices or nodes as formless place-holders, but as oscillators or dynamical systems coupled in the geometry of the network. Recent results for such situations have been developed and the study of dynamics on complex networks has begun.

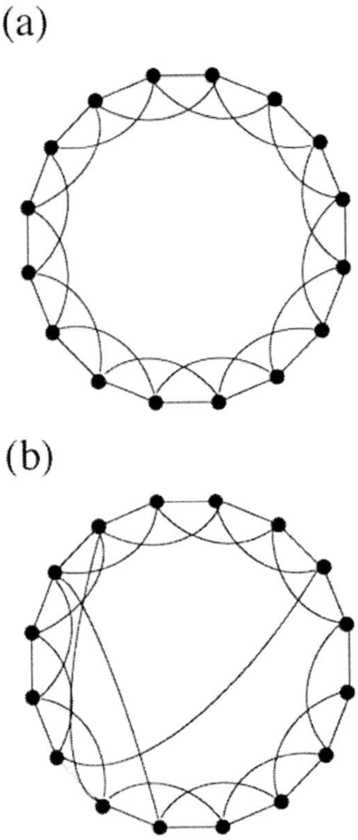

Fig. 1 Example of a cycle and a semiregular cycle (smallworld a la Watts&Strogatz)

In 1968 Watts and Strogatz [1] showed that simple cyclical networks called ***k*-cycles** (nodes connected to each other in circles, see Fig. 1 for an example) make the transition from networks where average distances between nodes is large to short average distance networks with the addition of surprisingly few edges randomly rearranged and reattached at random to other nodes in the network. At the same time the network remained highly **clustered** in the sense that nodes were connected in clumps. If we think of connected nodes as friends in a social network, highly clustered would mean that friends of a particular node would, with high probability be friends of each other. Thus with only a few percent or less of rearranged edges the network shrank in size, determined by average distance, but stayed localized in the clustering sense. These networks are referred to as **smallworld** networks. See Fig. 2 for a plot of fractional change in average distance and clustering vs. probability of edge rearrangement.

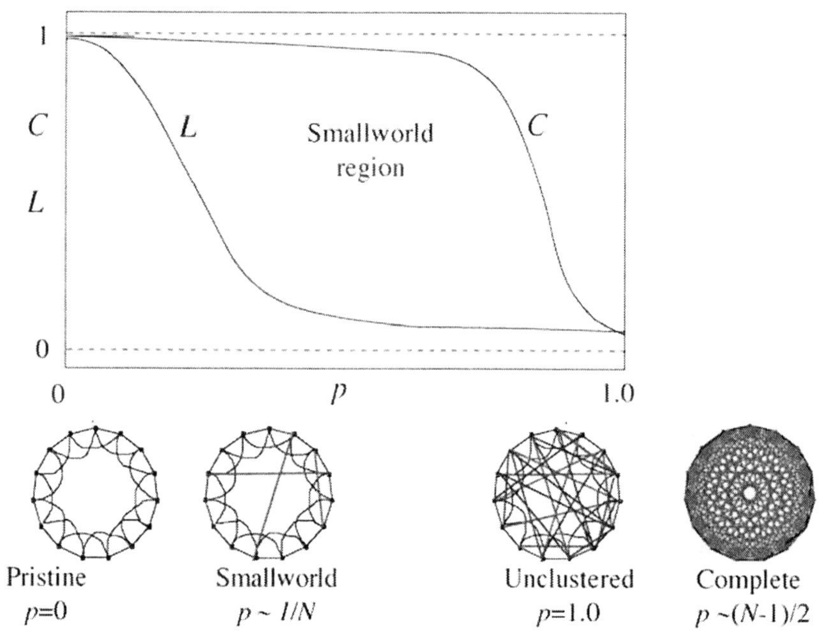

Fig. 2 Plot of L=L(p)/L(0) (normalized average distance between nodes) and C=C(p)/C(0) (normalized clustering) vs. p. Shown at the bottom are typical graphs that would obtain at the various p values including the complete graph. Note in the smallworld region we are very far from a complete graph.

Such smallworld networks are mostly regular with some randomness and can be referred to as semirandom. The number of edges connecting to each node, the **degree** of the node, is fairly uniform in the smallworld system. That is, the distribution of degrees in narrow, clustered around a well-defined mean. The Watts and Strogatz paper stimulated a large number of studies [2] and is seminal in opening interest in networks to new areas of science and with a new perspective on modeling actual networks realistically.

A little later in a series of papers Barabasi, Albert, and Jeong showed how to develop scale-free networks which closely matched real-world networks like co-authorship, protein-protein interactions and the world-wide web in structure. Such networks were grown a node at a time by adding a new node with a few edges connected to existing nodes. The important part of the construction was that the new node was connected to existing nodes with preferences for connections to those nodes already well-connected, i.e. with high degree. This type of construction led to a network with a few highly connected hubs and more lower connected hubs (see Fig. 3). In this network the **degree** of the node, has no well-defined average. The distribution of node degrees is a power law and there is no typical degree size in the sense that the degrees are not clustered around some mean value as in the smallworld case. The network is referred to as **scale-free**. The natural law of growth of the scale-free network, the rich get richer in a sense, seems to fit well into many situations in many fields. As a result interest is this network has grown quickly along with the cycle smallworld.

In the past decade in the field of nonlinear dynamics emphasis on coupled systems, especially coupled oscillators has grown greatly. One of the natural situations to study in arbitrarily connected identical oscillators is that of **complete synchronization** which in a discrete system is the analog of the uniform state in continuous systems like fluids where the uniform state would be laminar flow or chemical reactions like the BZ reaction where there is no spatial variation although temporal evolution can be very complex and/or chaotic. That is, in a completely synchronized system all the oscillators would be doing the same thing at the same time. The stability of the uniform state is of great interest for it portends the emergence of patterns when its stability is lost and it amounts to a coherent state when the stability can be maintained. The uniform or completely synchronized state is then a natural first choice to study in coupled systems.

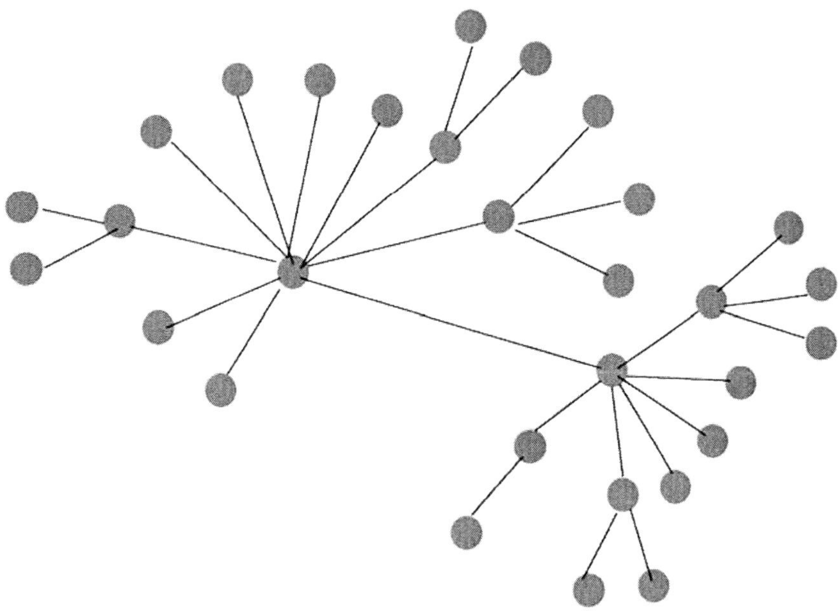

Fig. 3 Example of SFN with $m=1$. Note the hub structure.

In the last several years a general theory has been developed for the study of the stability of the synchronized state of identical oscillators in arbitrary coupling topologies [3,4]. A natural first step in the study of dynamics on complex or semirandom networks is the study of synchronization in cycle smallworlds and scale-free networks of oscillators. In the next section we develop the formalism of synchronization stability in arbitrary topologies and presents some ideas from networks and graph theory that will allow us to make some broad and generic conclusions.

II. Formal Development

A. Synchronization in Arbitrary, Coupled Systems

Here we present a formal development of the theory of stability of the synchronous state in any arbitrary network of oscillators. It is this theory which is very general that allows us to make broad statements about synchronous behavior in classes of semirandom networks.

We start with a theory based on linear coupling between the oscillators and show how this relates to an important quantity in the structure of a network. We then show how we can easily generalize the theory to a broader class of nonlinearly coupled networks of oscillators and iterated maps.

Let's start by assuming all oscillators are identical (that is, after all, how we can get identical or complete synchronization). This means the uncoupled oscillators have the following equation of motion,

$$\frac{d\mathbf{x}^i}{dt} = \mathbf{F}(\mathbf{x}^i) \tag{1}$$

where the superscript refers to the oscillator number (i=1,...,N) and subscripts on dynamical variables will refer to components of each oscillators, viz., x_j^i, j=1,...,m. We can linearly couple the N oscillators in some network by specifing a connection matrix, **G**, that consists of 1's and 0's to specify what oscillators is coupled to which other ones. We restrict our study to symmetric connections since our networks will have non-directional edges, hence, **G** is symmetric. Generalizations to non-symmetric couplings can be made (see Refs [4,5]).

We also assume all oscillators have an output function, **H**, that is a vector function of dimension m of the dynamical variables of each oscillator. Each oscillator has the same output function and its output is fed to other oscillators to which it is coupled. For example, **H**, might be an $m{\times}m$ matrix that only picks out one component to couple to the other oscillators.

The coupled equations of motion become [6],

$$\frac{d\mathbf{x}^i}{dt} = \mathbf{F}(\mathbf{x}^i) - \sigma\sum_{j=1}^{N} G_{ij}\mathbf{H}(\mathbf{x}^j), \tag{2}$$

where σ is the overall coupling strength and note that **G** acts on each oscillator as a whole and only determines which are connected and which are not. **H** determines which components are used in the connections. Since we want to examine the case of identical synchronization, we must have the equations of motion for all oscillators be the same when the system is synchronized. We can assure this by requiring that the sum $\sum_{j=1}^{N} G_{ij}\mathbf{H}(\mathbf{x}^j)$ is a constant when all oscillators are synchronous. The simplest constant is zero which can be assured by restricting the connection matrix **G** to have zero row sums. This works since all

$\mathbf{H}(\mathbf{x}^j)$ are them same at all times in the synchronous state. It means that when the oscillators are synchronized they execute the same motion as they do when uncoupled (Eq. (1)), except all variables are equal at all times. Generalization to non-zero constants can be done, but it unnecessarily complicates the analysis. A typical connection matrix is shown in the next equation,

$$\mathbf{G} = \begin{pmatrix} 2 & -1 & 0 & \ldots & 0 & -1 \\ -1 & 2 & -1 & 0 & \ldots & 0 \\ 0 & -1 & 2 & -1 & \ldots & 0 \\ & & \vdots & & & \\ 0 & \ldots & 0 & -1 & 2 & -1 \\ -1 & 0 & \ldots & 0 & -1 & 2 \end{pmatrix}, \tag{3}$$

for nearest neighbor, diffusive coupling on a ring or cycle.

Our central question is, for what types of oscillators (**F**), output functions (**H**), connection topologies (**G**), and coupling strengths (σ) is the synchronous state stable? Or more generally, for what classes of oscillators and networks can we get the oscillators to synchronize? The stability theory that emerges will allow us to answer these questions.

In the synchronous state all oscillators' variables are equal to the same dynamical variable: $\mathbf{x}^1(t) = \mathbf{x}^2(t) = \ldots = \mathbf{x}^N(t) = \mathbf{s}(t)$, where $\mathbf{s}(t)$ is a solution of Eq. (1). The subspace defined by the constraint of setting all oscillator vectors to the same, synchronous, vector is called the **synchronization manifold**. We test whether this state is stable by considering small, arbitrary perturbations ξ^j to each $\mathbf{x}^j$ and see whether all the perturbations ξ^j die out or grow. This is accompllished by generating an equation of motion for each ξ^j and determining a set of **Lyapunov exponents** which tell us the stability of the state. The use of Lyapunov exponents is the weakest condition for the stability of the synchronous state. Although other stability criteria can be used [5] we will use the Lyapunov exponents here.

To generate an equation of motion for the set of ξ^j we start with the full equations of motion for the network (Eq. (2)) and insert the perturbed value of the dynamical variables $\mathbf{x}^j(t) = \mathbf{s}(t) + \xi^j$ expanding all functions (**F** and **H**) in Taylor series to 1st order (we are only interested in small ξ^j values). This gives,

$$\frac{d\xi^i}{dt} = \sum_{j=1}^{N} \left[D\mathbf{F}(\mathbf{s})\delta_{ij} - \sigma G_{ij} D\mathbf{H}(\mathbf{s}) \right] \cdot \xi^j, \tag{4}$$

where $D\mathbf{F}$ and $D\mathbf{H}$ are the Jacobians of the vector field and the output function.

Eq. (4) is referred to as a **variational equation** and is often the starting point for stability determinations. This equation is rather complicated since given arbitrary coupling **G** it can be quite high dimensional. However, we can simplify the problem by noting that the equations are organized in block form. The blocks correspond to the (*ij*) indices of **G** and we can operating on them separately from the components within each block. We use this structure to diagonalize **G**. The first term with the Kronecker delta remains the same. This results in variational equations in eigenmode form:

$$\frac{d\zeta^l}{dt} = \left[D\mathbf{F}(\mathbf{s}) - \sigma \gamma_l D\mathbf{H}(\mathbf{s}) \right] \cdot \zeta^l, \tag{5}$$

where γ_l is the lth eigenvalue of **G**. We can now find the Lyapunov exponents of each eigenmode which corresponds to a "spatial" pattern of desynchronization amplitudes and phases of the oscillators. It would seem that if all the eigenmodes are stable (all Lyapunov exponents are negative), the synchronous state is stable, but as we will see this is not quite right and we can also simplify our analysis and not have to calculate the exponents of each eigenblock separately. We note that because of the zero-sum row constraint $\gamma=0$ is always an eigenvalue with eigenmode the major diagonal vector (1,1,...,1). We denote this as the first eigenvalue γ_1 since by design it is the smallest. The first eigenvalue is associated with the synchronous state and the Lyapunov exponents associated with it are those of the isolated oscillator. Its eigenmode represents perturbations that are the same for all oscillators and hence do not desynchronize the oscillators. The first eigenvalue is therefore not considered in the stability analysis of the whole system.

Next notice that Eq. (5) is the same form for all eigenmodes. Hence, if we solve a more generic variational equation for a range of couplings, then we can simply examine the exponents for each eigenvalue for stability. This is clearer if we show the equations. Consider the generic variational equation,

$$\frac{d\zeta}{dt} = \left[D\mathbf{F}(\mathbf{s}) - \alpha D\mathbf{H}(\mathbf{s}) \right] \cdot \zeta, \tag{6}$$

where α is a real number (**G** is symmetric and so has real eigenvalues). If we know the maximum Lyapunov exponent $\lambda_{max}(\alpha)$ for α over a range that includes the Lyapunov spectrum, then we automatically know the stability of all the modes by looking at the exponent value at each $\alpha=\sigma\gamma_l$ value. We refer to the function $\lambda_{max}(\alpha)$ as the **master stability function.**

For example, Fig. 4 shows an example of a typical stability curve plotting the maxium Lyapunov exponent vs. α. This particular curve would obtain for a particular choice of vector field (**F**) and output function (**H**). If the spectrum $\{\gamma_l\}$ all falls under the negative part of the stability curve (the deep well part), then all the modes are stable. In fact we need only look to see whether the largest γ_{max} and smallest γ_2, non-zero eigenvalues fall in this range. If there exists a continuous, negative λ_{max} regime in the stabilty diagram, say between α_1 and α_2, then it is sufficient to have the following inequality to know that we can always tune σ to place the entire spectrum of **G** in the negative area:

$$\frac{\gamma_{max}}{\gamma_2} < \frac{\alpha_2}{\alpha_1}, \tag{7}$$

We note two important facts: (1) We have reduced the stability problem for an oscillator with a particular stability curve (say, Fig. 4) to a simple calculation of the ratio of the extreme, non-zero eigenvalues of **G**; (2) Once we have the stability diagram for an oscillator and output function we do not have to re-calculate another stability curve if we reconfigure the network, i.e. construct a new **G**. We need only recalculate the largest and smallest, non-zero eigenvalues and consider their ratio again to check stability.

Finally, we remark that stability curves like Fig. 4 are quite common for many oscillators in the literature, especially those from the class arising from an unstable focus. Appendix A gives a heuristic reason for this common shape. For the rest of this article, we assume that we are dealing with the class of oscillators which have stability curves like Fig. 4. They may or may not be chaotic. If they are then at $\alpha=0$ $\lambda_{max} > 0$, otherwise $\lambda_{max} = 0$ at $\alpha=0$. And their values for large α may go positive or not for either chaotic or periodic cases. We will assume the most restrictive case that there is a finite interval where $\lambda_{max} < 0$ as in Fig. 4. This being the most conservative assumption will cover the largest class of oscillators including those which have multiple, disjoint α regions of stability as can happen in Turing pattern-generating instabilities [7]. Several other studies of the stability of the synchronous state have chosen weaker assumptions, including the assumption that the stability curve $\lambda_{max}(\alpha)$ becomes negative at some threshold (say, α_1) and remains negative for all $\alpha > \alpha_1$. Conclusions of stability in these cases only require the study of the first non-zero eigenvalue γ_2, but cover a smaller class of oscillators and are not as general as the broader assumption of Eq. (7).

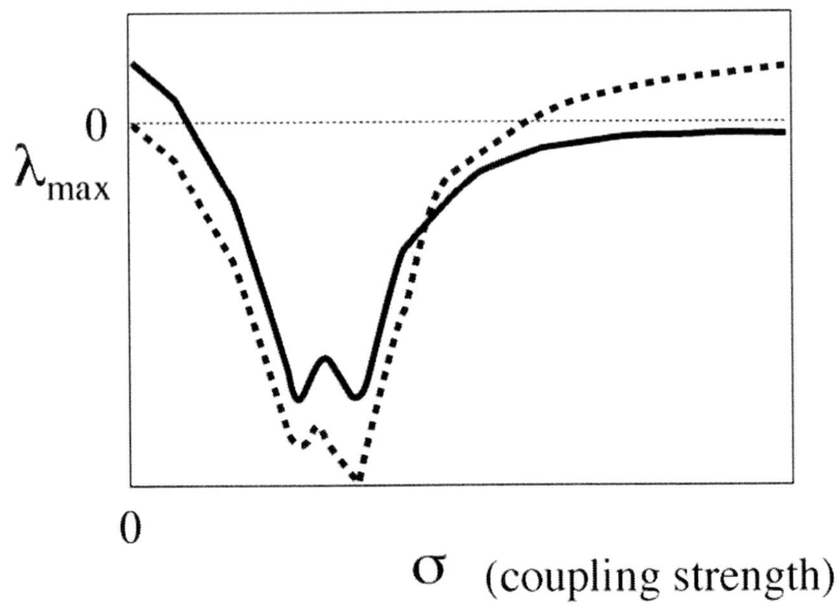

Fig. 4 Stability curve for a generic oscillator. The curve may start at (1) λ_{max}=0 (regular behavior), (2) λ_{max}>0 (chaotic behavior) or (3) λ_{max}<0 (stable fixed point) and asymptotically ($\sigma \rightarrow \infty$) go to (a) λ_{max}=0, (b) λ_{max}>0, (c) λ_{max}<0. Of course the behavior of λ_{max} at intermediate σ values is somewhat arbitrary, but typical stability curves for simple oscillators have a single minium. Shown are the combinations (1)-(b) and (2)-(c) for some generic simple oscillators.

B. Beyond Linear Coupling

We can easily generalize the above situation to one that includes the case of nonlinear coupling. If we write the dynamics for each oscillator as depending, somewhat aribtrarily on it's input from some other oscillators, then we will have the equation of motion,

$$\frac{d\mathbf{x}^i}{dt} = \mathbf{F}^i(\mathbf{x}^i, \mathbf{H}\{\mathbf{x}^j\}), \tag{8}$$

where here $\mathbf{F}^i$ is different for each i because it now contains arbitrary couplings. $\mathbf{F}^i$ takes N+1 arguments with $\mathbf{x}^i$ in the first slot and $\mathbf{H}\{\mathbf{x}^j\}$ in the remaining N slots. $\mathbf{H}\{\mathbf{x}^j\}$ is short for putting in N arguments which are the result of the output function $\mathbf{H}$ applied in sequence to all N oscillator vectors $\mathbf{x}^j$, viz., $\mathbf{H}\{\mathbf{x}^j\} = (\mathbf{H}(\mathbf{x}^1), \mathbf{H}(\mathbf{x}^2), \ldots \mathbf{H}(\mathbf{x}^N))$. We require the constraint,

$$\mathbf{F}^i(\mathbf{s}, \mathbf{H}\{\mathbf{s}\}) = \mathbf{F}^j(\mathbf{s}, \mathbf{H}\{\mathbf{s}\}), \tag{9}$$

for all i and j so that identical synchronization is possible.

The variational equation of Eq. (8) will be given by,

$$\frac{d\xi^i}{dt} = \sum_{j=1}^{N} \left[D_0 \mathbf{F}^i(\mathbf{s}, \mathbf{H}\{\mathbf{s}\})\delta_{ij} + D_j \mathbf{F}^i(\mathbf{s}, \mathbf{H}\{\mathbf{s}\}) \cdot D\mathbf{H}(\mathbf{s}) \right] \cdot \xi^j, \tag{9}$$

where D_0 is the partial derivative with respect to the first argument and D_j, j=1,2,...N, is the derivative with respect to the remaining N argument slots. Eq. (9) is almost in the same form as Eq. (4). We can regain Eq. (4) form by restricting our analysis to systems for which the partial derivatives of the second term act simply like weighting factors on the outputs from each oscillator. That is,

$$D_j \mathbf{F}^i(\mathbf{s}, \mathbf{H}\{\mathbf{s}\}) = -\sigma G_{ij} \mathbf{1}_m, \tag{10}$$

where G_{ij} is a constant and $\mathbf{1}_m$ is the $m \times m$ unit matrix. Now we have recovered Eq. (4) exactly and all the analysis that led up to the synchronization criterion of Eq. (7) applies. Note that Eq. (10) need only hold on the synchronization manifold. Hence, we can use many forms of nonlinear coupling through the $\mathbf{F}^i$ and/or $\mathbf{H}$ functions and still use stability diagrams and Eq. (7).

C. Networks and Graph Theory

A **network** is synonomous with the definition of a **graph** and when the nodes and/or edges take on more meaning like oscillators and couplings, respectively, then we should properly call the structure a network of oscillators, etc. However, we will just say network here without confusion. Now, what is a graph? A graph U is a collection of **nodes** or **vertices** (generally, structureless entities, but oscillators herein) and a set of **connections** or **edges** or **links** between some of them. See Fig. 1. The collection of vertices (nodes) are usually denoted as $V(U)$ and the collection of edges (links) as $E(U)$. We let N denote the number of vertices, the **cardinality** of U written as $|U|$. The number of edges can vary between 0 (no vertices are connected) and $N(N-1)/2$ where every vertex is connected to every other one.

The association of the synchronization problem with graph theory comes through a matrix that appears in the variational equations and in the analysis of graphs. This is the connection matrix or **G**. In graph theory it is called the **Laplacian** since in many cases like Eq. (3) it is the discrete version of the second derivative $\Delta=\nabla^2$. The Laplacian can be shown to be related to some other matrices from graph theory.

We start with the matrix most studied in graph theory, the **adjacency** matrix **A**. It is given by the symmetrix form where $A_{ij}=1$ if vertices i and j are connected by an edge and $A_{ij}=0$ otherwise. For example, for the graph in Fig. 5 has the following adjacency matrix,

$$\mathbf{A}=\begin{pmatrix} 0 & 1 & 1 & 0 & 0 \\ 1 & 0 & 1 & 0 & 1 \\ 1 & 1 & 0 & 1 & 0 \\ 0 & 0 & 1 & 0 & 0 \\ 0 & 1 & 0 & 0 & 0 \end{pmatrix}, \tag{11}$$

Much effort in the mathematics of graph theory has been expended on studying the adjacency matrix. We will not cover much here, but point out a few things and then use **A** as a building block for the Laplacian, **G**.

The components of the powers of **A** describe the number of steps or links between any two nodes. Thus, the non-zero components of $\mathbf{A}^2$ show which nodes are connected by following exactly two links (including traversing back over the same link). In general, the mth power of **A** is the matrix whose non-zero components show which nodes are connected by m steps. Note, if after N–1th step **A** still has a zero, off diagonal component, then the graph must be **disconnected**. That is, it can be split into two subgraphs each of whose nodes have no edges connecting them to the other subgraph. Given these minor observations and the fact that a matrix must satisify its own characteristic equation, one can appreciate that much work in graph theory has gone into the study of the eigenspectrum of the adjacency matrix. To pursue this further, we recommend the book Ref. [8].

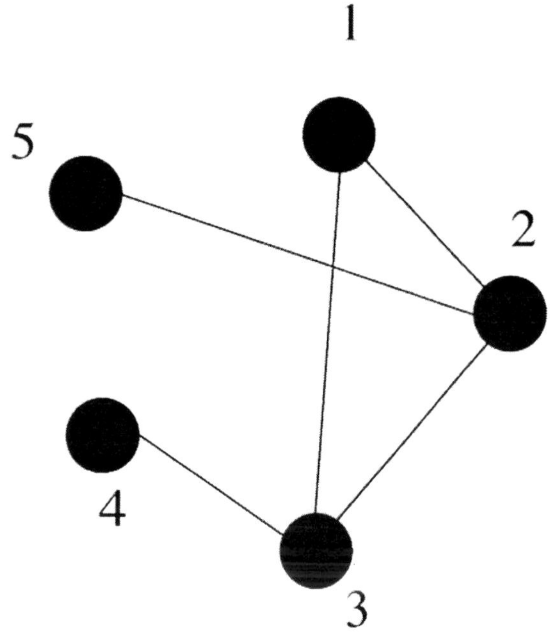

Fig. 5 Simple graph generating the adjacency matrix in Eq. (11)

The **degree** of a node is the sum of the number of edges connecting it to other nodes. Thus, in Fig. 5, node 2 has a degree = 3. We see that the degree of a node is just the row sum of the row of **A** associated with that node. We form the degree matrix or **valency** matrix **D** which is a diagonal matrix whose diagonal entries are the row sums of the corresponding row of **A**, viz., $D_{ij}=\Sigma_k A_{ik}$. We now form the new matrix, the Laplacian **G**=**D**–**A**. For example, for the diffusive coupling of Eq. (3), **A** would be a matrix like **G**, but with 0 replacing 2 on the diagonal and **D** would be the diagonal matrix with 2's on the diagonals. The eigenvalues and vectors of **G** are also studied in graph theory [8,9], although not as much as the adjacency matrix. We have seen how **G**'s eigenvalues affect the stability of the synchronous state so some results of graph theory on the eigenspectrum of **G** may be of interest and we present several below.

The Laplacian is a postive, semi-definite matrix. We assume its N eigenvalues are ordered as $\gamma_1<\gamma_2<...<\gamma_N$. The Laplacian always has 0 for the smallest eigenvalue associated with the eigenvector $\mathbf{v}_1$=(1,1,...,1), the major diagonal. The latter is easy to see since **G** must have zero row sum by construction. The next largest eigenvalue, γ_2 is called the **algebraic connectivity**. To see why, consider a graph which can be cleanly divided into two disconnected subgraphs. This means by rearranging the numbering of the nodes **G** would be divided into two blocks with no non-zero "off-diagonal" elements since they are not connected,

$$\mathbf{G} = \begin{pmatrix} \mathbf{G}_1 & 0 \\ 0 & \mathbf{G}_2 \end{pmatrix}, \tag{12}$$

where we assume $\mathbf{G}_1$ is $n_1 \times n_1$ dimensional and $\mathbf{G}_2$ is $n_2 \times n_2$ dimensional. In this latter case we now have two zero eigenvalues each associated with unique eigenvectors mutually orthogonal, namely, $\mathbf{v}_1$=(1,1,...,1,0,0,...,0) with n_1 1's and $\mathbf{v}_2$=(0,0,...,0,1,1,...,1) with n_2 1's. The degeneracy of the zero eigenvector is one greater than the connectivity. If $\gamma_2>0$, the graph is connected. We immediately see that this has consequences in synchronization stabilty, since a disconnected set of oscillators cannot synchronize physically and mathematically this shows up as a zero eigenvector.

Much work has gone into obtaining bounds on eigenvalues for graphs [8,9]. We will present some of these bounds (actually deriving a few ourselves) and show how they can lead to some insight into synchronization conditions in general. In the following the major diagonal vector (1,1,...,1) which is the eigenvector of γ_1 is denoted by $\boldsymbol{\theta}$.

We start with s few well-known min-max expressions for eigenvalues of a real symmetric matrix which is **G** in our case, namely,

$$\gamma_1 = \min\left\{ \frac{\langle \mathbf{G}\mathbf{v}, \mathbf{v} \rangle}{\langle \mathbf{v}, \mathbf{v} \rangle} \middle| \mathbf{v} \neq 0, \mathbf{v} \in \mathbf{R}^N \right\}, \tag{13}$$

$$\gamma_2 = \min\left\{ \frac{\langle \mathbf{G}\mathbf{v}, \mathbf{v} \rangle}{\langle \mathbf{v}, \mathbf{v} \rangle} \middle| \mathbf{v} \neq 0, \mathbf{v} \in \mathbf{R}^N, \mathbf{v} \perp \boldsymbol{\theta} \right\}, \tag{14}$$

and,

$$\gamma_{\max} = \max\left\{ \frac{\langle \mathbf{G}\mathbf{v}, \mathbf{v} \rangle}{\langle \mathbf{v}, \mathbf{v} \rangle} \middle| \mathbf{v} \neq 0, \mathbf{v} \in \mathbf{R}^N \right\}, \tag{15}$$

Now with the proof of two simple formulas we can start deriving some inequalities.

First we show that $\sum_{i,j=1}^{N} A_{ij}\left(v_i - v_j\right)^2 = 2\langle \mathbf{Gv}, \mathbf{v}\rangle$, where A_{ij} are the components of the adjacency matrix. By symmetry of **A**, we have

$$\sum_{i,j=1}^{N} A_{ij}\left(v_i - v_j\right)^2 = 2\sum_{i,j=1}^{N} (A_{ij}v_j^2 - A_{ij}v_j v_i),$$

The first term in the sum is the row sum of **A** which is the degree D_{ii} making the term in parenthesis an matrix product with the Laplacian so that we have $\sum_{i,j=1}^{N} v_i\left(D_{ii}\delta_{ij} - A_{ij}\right)v_j = 2\langle \mathbf{Gv}, \mathbf{v}\rangle$, thus proving the first formula.

Second, we show that $\sum_{i,j=1}^{N} \left(v_i - v_j\right)^2 = 2N\langle \mathbf{v}, \mathbf{v}\rangle$. Because we are using $\mathbf{v} \perp \boldsymbol{\theta}$ we can easily show that $\sum_{i,j=1}^{N} \left(v_i - v_j\right)^2 = 2\sum_{i,j=1}^{N} \left(v_i^2 - v_i v_j\right) = 2\sum_{i,j=1}^{N} \left(v_i^2 - \langle \mathbf{v}.\boldsymbol{\theta}\rangle^2\right) = 2N\langle \mathbf{v}, \mathbf{v}\rangle$ since $\mathbf{v} \perp \boldsymbol{\theta}$.

Noting that the $\left(v_i - v_j\right)$ terms are not affected if we add a constant to each component of **v**, that is, $\left(v_i - v_j\right) = \left((v_i + c) - (v_j + c)\right)$. We can now write Eq.(14) and Eq. (15) as

$$\gamma_2 = N \min\left\{ \frac{\sum_{i,j=1}^{N} A_{ij}(v_i - v_j)^2}{\sum_{i,j=1}^{N} (v_i - v_j)^2} \middle| \mathbf{v} \in \mathbf{R}^N, \mathbf{v} \neq c\boldsymbol{\theta}, c \in \mathbf{R} \right\}, \tag{16}$$

and,

$$\gamma_{\max} = N \max\left\{ \frac{\sum_{i,j=1}^{N} A_{ij}(v_i - v_j)^2}{\sum_{i,j=1}^{N} (v_i - v_j)^2} \middle| \mathbf{v} \in \mathbf{R}^N, \mathbf{v} \neq c\boldsymbol{\theta}, c \in \mathbf{R} \right\}, \tag{17}$$

If δ is the minimum degree and Δ the maximum degree of the graph then the following inqualities can be derived:

$$\gamma_2 \leq \frac{N}{N-1}\delta \leq \frac{N}{N-1}\Delta \leq \gamma_{\max} \leq \max\{D_{ii}+D_{jj}\} \leq 2\Delta, \tag{18}$$

where $D_{ii}+D_{jj}$ is the sum of degrees of two nodes that are connected by an edge. We can prove the first inequality by choosing a particular **v**, namely the standard basis $\mathbf{e}^{(i)}$= all zeros except for a 1 in the ith position. Putting $\mathbf{e}^{(i)}$ into Eq. (16) we get $\gamma_2 \leq \frac{ND_{ii}}{N-1}$. This last relationship is true for all D_{ii} and so is true for the minimum, δ. A similar argument gives the first $\gamma_{\max}$ -Δ inequality. The remaining inequalities bounding $\gamma_{\max}$ from above are obtained through the Perron-Frobenius theorem [9].

There remains another inequality that we will use, but not derive here. It is the following:

$$\gamma_2 \geq \frac{4}{N\,\mathrm{diam}(U)}, \tag{19}$$

where diam(U) is the diameter of the graph. The distance between any two nodes is the minimum number of edges transversed to get from one node to the other. The diameter is the maximum of the distances [8]. There are other inequalities, but we will only use what is presented up to here.

The synchronization criterion is that the ratio $\gamma_{\max}/\gamma_2$ has to be less than some value (α_2/α_1) for a particular oscillator node. We can see how this ratio will scale and how much we can bound it using the inequalities. First, we note that the best we can do is $\gamma_{\max}/\gamma_2$=1 which occurs only when each node is connected to all other nodes. We can estimate how much we are bounded away from 1 by using the first inequalities of Eq. (18) . This gives,

$$\frac{\Delta}{\delta} \leq \frac{\gamma_{\max}}{\gamma_2}, \tag{20}$$

Hence, if the degree distributions are not narrow, that is, there is a large range of distributions, then the essential eigenratio $\gamma_{\max}/\gamma_2$ cannot be close to 1 possibly making the oscillators difficult to synchronize (depending on α_2/α_1). Note that this does not mean necessarily that if δ/Δ is near 1 we have good synchronization. We will see below a case where δ/Δ=1, but γ_{max}/γ_2 is large and not near 1.

To have any hope in forcing the eigenratio down we need an upper bound. We combine Eq. (18) with Eq. (19) . This gives,

$$\frac{\gamma_{max}}{\gamma_2} \leq \frac{N \,\mathrm{diam}(U) \max\{D_{ii} + D_{jj} \mid i \text{ and } j \text{ are connected}\}}{4}, \qquad (21)$$

This inequality is not very strong since even if the network has small diameter and small degrees the eigenratio still scales as N. We want to consider large networks and obviously this inequality will not limit the eigenratio as the network grows. However, it does suggest that if we can keep the degrees low and lower the diameter we can go from a high-diameter, possibly hard to synchronize network to an easier to synchronize one. We will see this is what happens in smallworld systems.

III. Synchronization in Semirandom Smallworlds (Cycles)

A. Generation of Smallworld Cycles

The generation of smallworld cycles starts with a k-cycle (defined shortly) and then either rewires some of the connections randomly or adds new connections randomly. We have chosen the case of adding more, random connections since we can explore the cases from pristine k-cycle networks all the way to compete networks (all nodes connected to all other nodes).

To make a k-cycle we arrange N nodes in a ring and add k connections to each of the k nearest neighbors to the right. This gives a total of kN edges or connections. Fig. 1 shows a 2-cycle. We can continue the construction past this initial point by choosing a probability p and going around the cycle and at each node choosing a random number r in (0,1) and if $r \leq p$ we add an edge from the current node to a randomly chosen node currently not connected to our starting node. Note that we can guarantee adding an edge to each node by choosing p=1. We can extend this notion so we can reach a complete neetwork by allowing probabilities greater than 1 and letting the integer part specify how many times we go around the cycle adding edges with probability p=1 with the final loop around the cycle using the remainig fraction of probability. For example, if we choose

p=2.4 we go twice around the cycle adding an edge at each node (randomly to other nodes) and then a third time around adding edges with probability 0.4. With this contruction we would use a probability $p=N(N-2k-1)/2$ to make a complete network.

We refer to the k-cycle without any extra, random edges as the pristine network. It is interesting to examine the eigenvalues of the pristine Laplacian especially as they scale in ratio.

B. Eigenvalues of Pristine Cycles

The Laplacian for the pristine k-cycle is **shift invariant** or **circulant** which means that the eigenvalues can be calculated from a discrete Fourier transform of a row of the Laplacian matrix. This action gives for the lth eigenvalue,

$$\gamma_l = 2\left[k - \sum_{j=1}^{k} \cos\left(\frac{2\pi(l-1)j}{N}\right)\right]. \tag{22}$$

What we are mostly interested in here is the eigenratio γ_{max}/γ_2. We might even question whether this ratio is sufficiently small for the pristine lattices that we might not even need extra, random connections. Fig. 7 shows the eigenratios as a function of N for k=1, 2, 3 and 4. The log-log axes of Fig. 7 shows that γ_2 scales roughly as $(2k^3+3k^2+k)/N^2$, γ_{max} values are constant for each k, and, therefore, the eigenratio γ_{max}/γ_2 scales as $N^2/(2k^3+3k^2+k)$. The scaling of the eigenratio is particularly bad in that the ratio gets large quickly with N and given that for many oscillators the stability region (α_2/α_1) is between 10 and a few hundred we see that pristine k-cycles will generally not produce stable synchronous behavior. Hence, we must add random connections in the hope of reducing γ_{max}/γ_2.

C. Eigenvalues of Smallworld Cycles

In this section and the next on SFNs we look at the trends in eigenratios as new edges are randomly added increasing the coupling between oscillators. If these new edges are added easily in a system, then surely the best choice is to form the complete network and have the best situation possible, $\gamma_{max}/\gamma_2=1$. However, we take the view that in many systems there is some expense or cost in adding extra edges to pristine networks. For biological systems the cost will be in extra intake of food and extra energy which will be fruitful only if the new networks are a great improvement and provide the organisms some evolutionarly advantages. In human-engineered projects, the cost is in materials, time, and perhaps maintenance. Therefore, the trend in γ_{max}/γ_2 as edges are added randomly will be considered important.

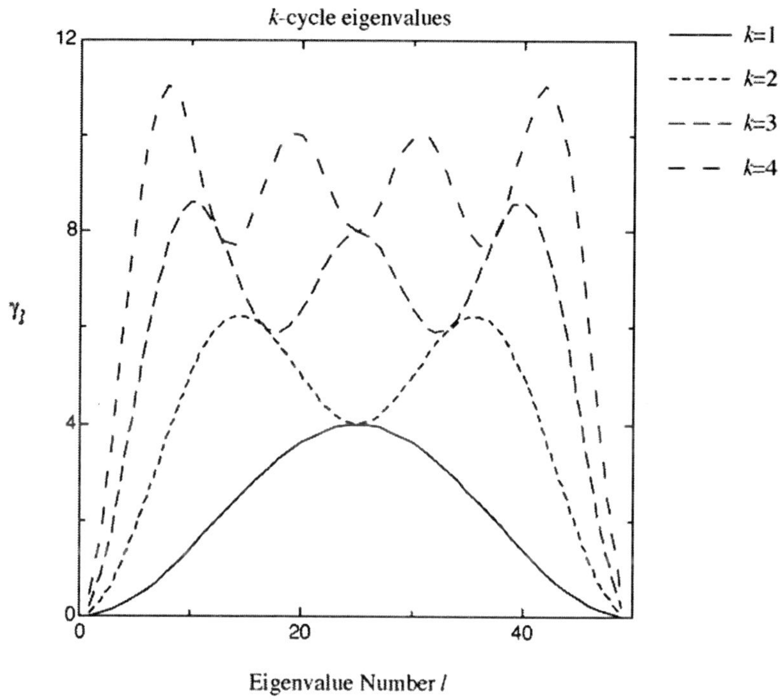

Fig. 6 Eigenvalues of the pristine k-cycle for k=1, 2, 3, and 4

Fig. 8 shows the eigenratio for k-cycle networks as a function of f the fraction of the complete lattice that obtains when edges are added randomly for probabilities from p=0 to values of p yielding complete graphs for N=50 nodes. Fig. 9 shows the same for 100 nodes. For now we ignore the plots for SFNs and concentrate only on the trends in the k-cycle, semirandom networks.

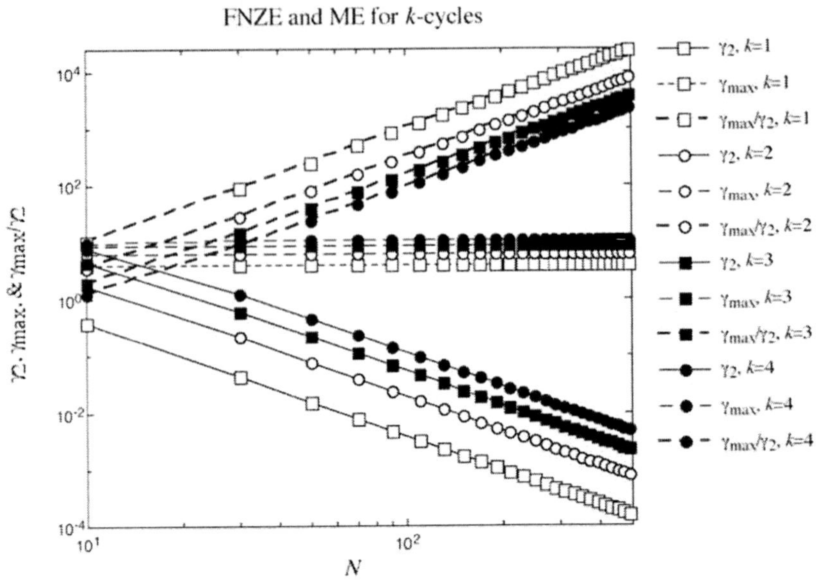

Fig. 7 Eigenratios of the pristine k-cycle as a function of N for k=1, 2, 3 and 4.

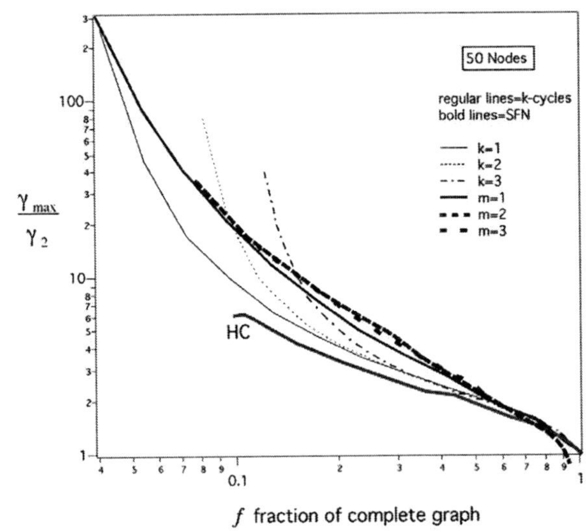

Fig. 8 Eigenratio for 50 node k-cycle semirandom network as a function of fraction of complete network.

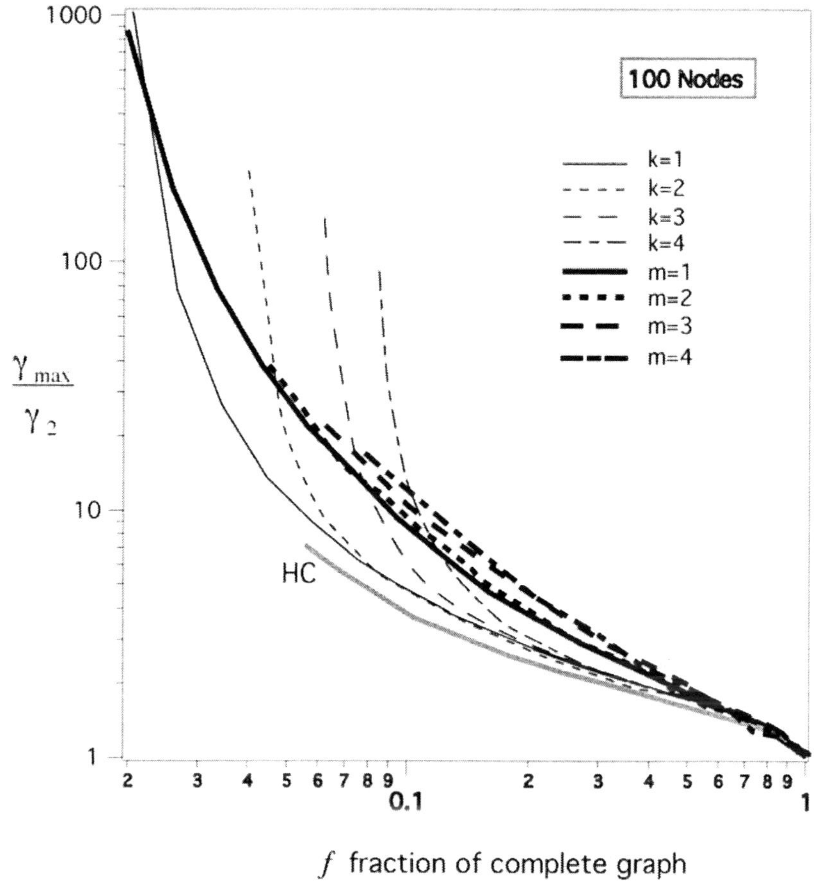

Fig. 9 Eigenratio for 100 node *k*-cycle semirandom network as a function of fraction of complete network.

We saw in the previous section that the worse case for pristine *k*-cycles is when *k*=1. However, Fig. 8 and Fig. 9 show that this turns into the best case of all the *k*-cycles when random edges are added if we compare the networks at the same fraction of complete graph values where the total number of edges or cost to the system is the same. Although each case for *k*>1 starts at lower γ_{max}/γ_2 values for the pristine lattices, at the same fraction of complete graph values the *k*=1 network has lower values. This immediately suggests that a good way to generate a synchronizable network is to start with a *k*=1 network and randomly add edges. We will compare this to starting with other networks below.

In terms of graph diameter we can heuristically understand the lowering of γ_{max}/γ_2. Recall inequality Eq. (21). This strongly suggests that as diam(U) decreases the eigenratio must decrease. From the original Watts and Strogatz paper we know the average distance between nodes decreases early upon addition a few random edges. Although average distance is not the same as diameter, we suspect that since the added edges are randomly connected the diameter also decreases rapidly. Note that the k-cycles are well-behaved in regard to the other inequality Eq. (20). In fact δ/Δ starts out with a value of 1 in the pristine lattices. Of course, the eigenratio is not forced to this value from above since the diameter is large for the pristine networks – on the order of $N/(4k)$ which means the upper bound grows as N^2 just the same trend as the actual eigenratio for the pristine networks. However, adding edges does not change δ/Δ by much since the additions are random, but it does force the upper bound down.

IV. Synchronization in Scale-Free Networks

A. Generation of Scale-Free Networks

Scale-free networks (SFNs) get their name because they have no "typical" or average degree for a node. The distribution of degrees follows an inverse power law first discovered by Albert and Barabási [10] who later showed that one could derive the power law in the "continuum" limit of large N. Several methods have been given to generate SFNs, but here we use the original method [11].

The SFN is generated by an iterative process. We start with an initial number of nodes connected by single edges. The actual initial state does not affect the final network topology provided the initial state is a very small part of the final network. We now add nodes one at a time and each new node is connected to m existing nodes with no more than one connection to each (obviously the initial number of nodes must be equal to or greater than m). We do this until we have the desired number of nodes, N. The crucial part is in how we choose the m nodes to connect each new node to. This is done by giving each existing node a probability P_i which is proportional to the degree of that node. Normalizing the probabilities gives,

$$P_i = \frac{d_i}{\sum_{j \in \substack{\text{existing} \\ \text{nodes}}} d_j}, \tag{23}$$

where d_j is the degree of the jth node. Using this probability we can form the cumulative set of intervals, $\{P_1, P_1+P_2, P_1+P_2+P_3,...\}$. To choose an existing node for connection we randomly pick a number from the interval (0,1), say c, and see which interval it falls into. Then we pick the node of that interval with the highest index. A little thought will show that the ordering of the nodes in the cumulative intervals will not matter since c is uniformly distributed over (0,1).

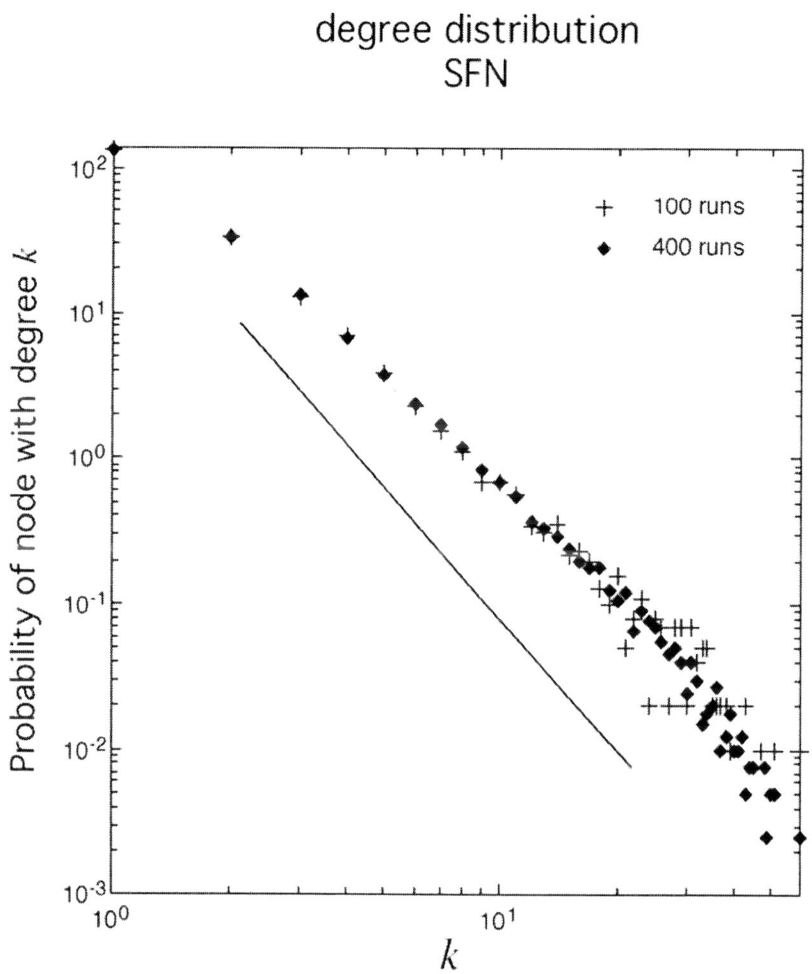

Fig. 10 Distribution of node degrees for SFNs.

Such a network building process is often referred to as "the rich get richer." It is easy to see that nodes with larger degrees will be chosen preferentially. The above process forms a network in which a few nodes form highly connected hubs, more nodes form smaller hubs, etc. down to many single nodes connected only through their original *m* edges to the network. This last sentence can be quantified by plotting the distribution or count of the degrees vs. the degree values. This is shown in Fig. 10.

We note that since we are judging synchronization efficiency of networks we want to compare different networks (e.g. cycles and SFNs) which have roughly the same number of edges, where we view adding edges as adding to the "cost" of building a network. We can easily compare pristine cycles and SFNs since for $k=m$ we have almost the exact same number of edges (excepting the initial nodes of the SFN) for a given N.

B. Eigenvalues of Pristine Scale-Free Networks

Although we do not start with a fully deterministic network in the SFN case, we can still characterize the eigenvalues and, to a lesser extenct, the eigenvectors of the original SFN. This is useful when comparing the effect of adding more, randomly chosen edges as we do in the next section.

Fig. 11 shows the FNZE, ME, and their ratio vs. N for various m values of 1, 2, 3, and 4. For $m=1$ we have (empirically) $\gamma_{max}\sim\sqrt{N}$ and $\gamma_2\sim 1/N$. Therefore the ratio scales as $\gamma_{max}/\gamma_2\sim N^{\frac{3}{2}}$. Recall that the pristine cycle's ratio scaled as $\gamma_{max}/\gamma_2\sim N^2$, hence for the same number of nodes and $k=m$ the pristine eigenratio of cycles grows at a rate $\sim\sqrt{N}$ faster then the SFN eigenratio. Thus, for large networks it would seem that SFNs would be the network of choice compared to cycles for obtaining the generic synchronization condition $\gamma_{max}/\gamma_2\leq\alpha_2/\alpha_1$ for the same number of edges. However, as we add random edges we see that this situation does not maintain. For values of $m\geq 2$, the situation is not as clear.

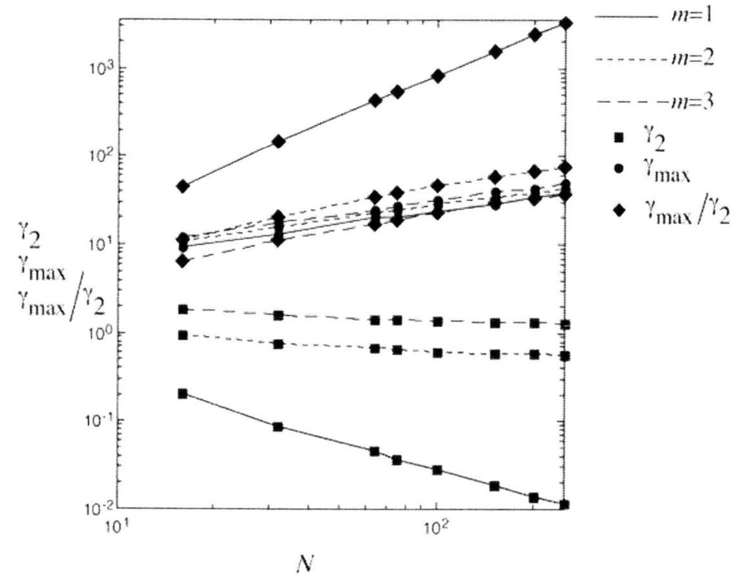

Fig. 11 FNZE, ME, and their ratio vs. N for SFNs.

C.. Eigenvalues of Scale-Free Networks

Fig. 8 and Fig. 9 also show the eigenratio for SFNs as a function of f the fraction of complete graph. Note that we can directly compare SW and SFNs when $m=k$ for which they have the same f value. We are now in a position to make several conclusions regarding the comparison of SW networks and SFNs. Such comparisons seem to hold across several values of N. We tested cases for N=50, 100, 150, 200, and 300.

First SW semirandom k cycles start out in their pristine state with a larger eigenratio than SFNs with $m=k$. However, with the addition of a small number of edges – barely changing f the SW k cycle eigenratio soon falls below its SFN counterpart with $m=k$. The eigenratios for all SFNs fall rapidly with increasing f, but they do not reach the same low level as the SW networks until around f=0.5. This implies that the SWs are more efficient than SFNs in reducing eigenratio or, equivalently, synchronizing oscillators.

An interesting phenomenon for SFNs is that changing m only appears to move the eigenratio up or down the m=1 curve. That is, curves for SFNs with different m values fall on top of each other with their starting point being higher f

values for higher m values. This suggests an underlying universal curve which the m=1 curve is close to. The connection to the previous section where level spacing for the SFN eigenvalues begins to appear random-like as we go to higher m values is mirrored in the present section by showing that higher m values just move the network further along the curve into the randomly added edge terroritory.

At f=1 all networks are the same. They are complete networks with $N(N-1)/2$ edges. At this point the eigenratio is 1. As we move back away from the f=1 point by removing edges, but preserving the underlying pristine network, the networks follow a "universal" curve down to about f=0.5. Larger differences between the networks don't show up until over 50% of the edges have been removed which implies that the underlying pristine skeleton does not control the eigenratio beyond f=0.5. This all suggests that an analytic treatment of networks beyond f=0.5 might be possible. We are investigating this.

In their original paper Watts and Strogatz [12] suggested that the diminishing average distances would allow a smallworld network of oscillators to couple more efficiently to each other and therefore synchronize more readily. In Fig. 12 we plot the average distance in each network as a function of fraction of completed graph f. In Fig. 13 the same type of plot is given for the clustering coefficient. What we immediately notice is that there seem to be no obvious relationship to eigenratios as a function of f. SFNs start out with very small average distance and have very low clustering coefficients, at least until many random edges are added to beyond f=0.1 which is beyond the smallworld regime. Thus, it appears that neither average distance nor clustering affect the eigenratio. What network statistics, if any, do affect γ_{max}/γ_2?

From the graph theory bounds in section II.C. Eqs. (20) and (21) we can explain some of the phenomena we find numerically for eigenratios. We can immediately explain the higher eigenratio of SFNs using Eq. (20). The eigenratio is bounded away from 1 (the best case) by the ratio of largest to smallest degrees (Δ/δ). For smallworld k-cycles this ratio starts at 1 (since all nodes have degree $2k$) and does not change much as edges are added randomly. Hence, it is at least possible for the eigenratio to approach 1, although this argument does not require it. Conversely, in the SFN the degree distribution is wide with many nodes with degree=m and other nodes with degree on the order of some power of N. Hence, for SFNs Δ/δ is large, precluding the possibility that γ_{max}/γ_2 can get close to 1 until many random edges are added.

We can bound the eigenratio from above using Eq. (21). Thus, γ_{max}/γ_2 is sandwiched between Δ/δ and $\frac{1}{4} N\,\mathrm{diam}(U)\max\{D_{ii}+D_{jj}\}$. We show this schematically in Fig. 14. The ratio Δ/δ provides a good lower bound, but the upperbound of Eq. (21) is not tight in that it scales with *N*. Hence, even in small networks the upper bound is an order of magnitude or more larger than the good lower bound. At this stage it does not appear that graph theory can give the full answer as to why smallworld *k*-cycles are so efficient for synchronizing oscillators.

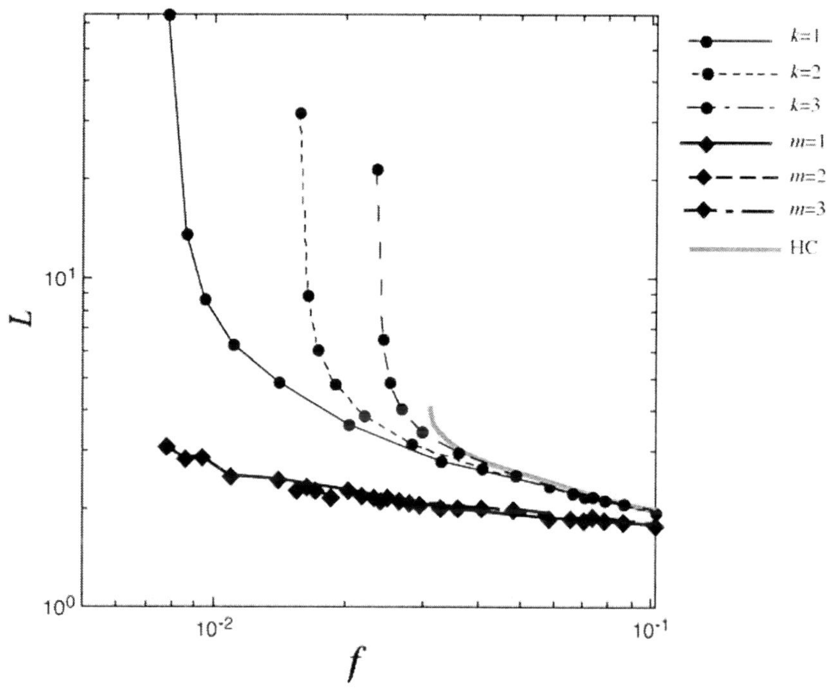

Fig. 12 Average distances between nodes on SW, SFN, and HC networks.

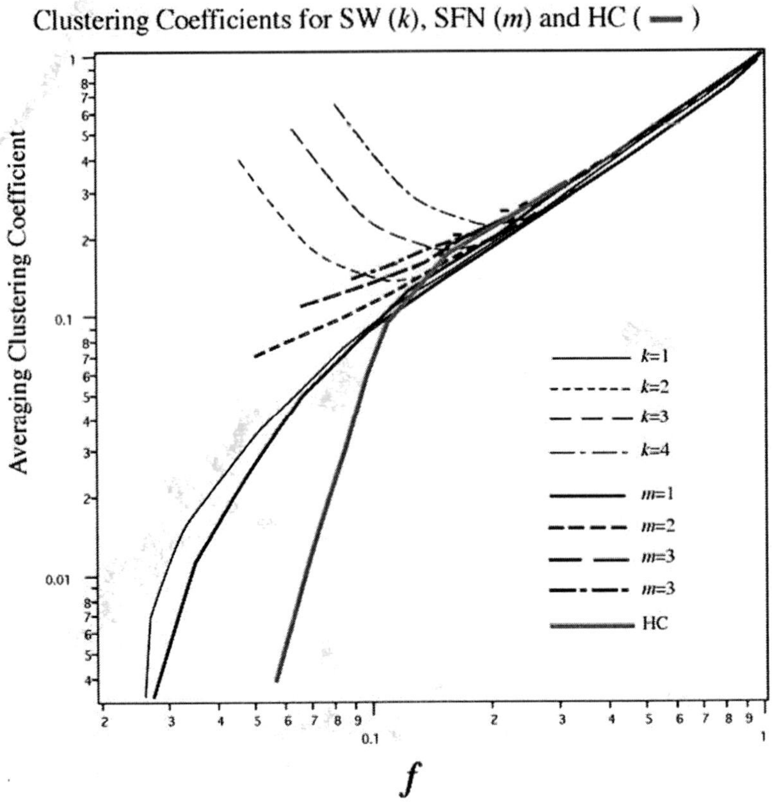

Fig. 13 Clustering coefficient for SW, SFN, and HC networks.

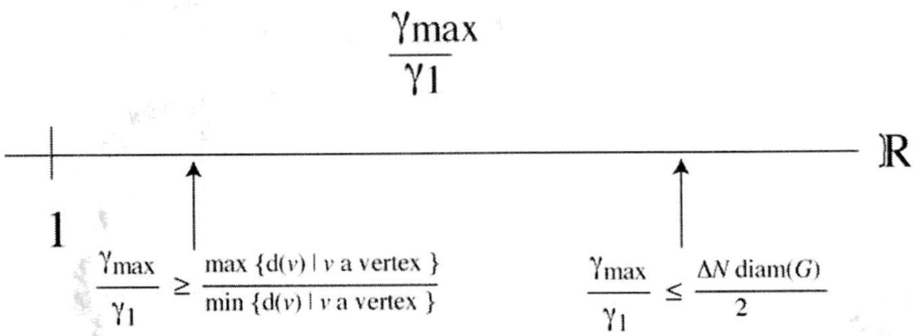

Fig. 14 The bounded interval containing the eigenratio γ_{max}/γ_2.

V. HYPERCUBE NETWORKS

As a simple third comparison we examine the eigenratio of the **hypercube** (HC) network which we showed in Ref. [13] was a very efficient network for obtaining low eigenratios. HC networks are built by connecting $N=2^D$ nodes in a topology so that all nodes form the corners of a D-dimensional hypercube with degree D. Hence, the 1st graph theory inequality (Eq. (20)) allows the eigenratio to be the minimum allowed value of 1 although this is not mandatory. Furthermore, it is not hard to show that the maximum eigenvalue scales as $2D=2\log_2(N)$ and the smallest eigenvalue is always 2 so that $\gamma_{max}/\gamma_2=D$ for the pristine HC network and this increases very slowly with N. We therefore expect the eigenratio for any HC network to start off small and decrease to 1.

Fig. 8 and Fig. 9 contain plots of the HC network eigenratio vs. f from pristine state to complete network. Indeed, the initial eigenratios are low, but this comes with a price. The price contains two contributions. One is that we need to start at a much higher f value, that is, HC networks in their pristine state require more edges than the SW $k=1$ network. And the other contribution is that the HC must be constructed carefully since the pristine state is apparently more complex than any other network.

In the end one gets the same performance by just connecting all the nodes in a loop and then adding a small number of random edges. The construction is simpler and the synchronization as good as one of the best networks, the HC.

V. CONCLUSIONS

Our earlier work [13] showed that many regular networks were not as efficient in obtaining synchronization as the smallworld network or fully random networks. Here we added two other semirandom networks, the SFN+random edges and the HC+random edges to our analysis. The results appear to be that elaborately constructed networks offer no advantage and perhaps even disadvantages over the simple k-cycle+random edges, especially in the smallworld regime.

The fully random network [14] can also have ratios γ_{max}/γ_2 which are similar to the smallworld case for large enough probabilities for adding edges. However, the random networks are never guaranteed to be fully connected. When unconnected, as we noted earlier, synchronization is impossible. The observation we can make then is that the 1-cycle smallworld network may be the best compromise between fully random and fully regular networks. The smallworld network maintains the fully connectedness of the regular networks, but gains all the advantages of the random networks for efficient synchronization.

REFERENCES

[1] D.J. Watts, *Small Worlds*. (Princeton University Press, Princeton, NJ, 1999).

[2] M.E.J. Newman and D.J. Watts, *Physics Letters* **A 263** (4-6), 341 (1999); M.E.J. Newman, C. Moore, and D.J. Watts, *Physical Review Letters* **84** (14), 3201 (2000); C. Moore and M.E.J. Newman, *Physical Review* **E 61** (5), 5678 (2000); L.F. Lago-Fernandez, R. Huerta, F. Corbacho et al., *Physical Review Letters* **84** (12), 2758 (2000); Rémi Monasson, *European-Physical-Journal-B* **12** (4), 555 (1999); R.V. Kulkami, E. Almaas, and D. Stroud, *Physical Review* **E 61** (4), 4268 (2000); M. Kuperman and G. Abramson, *Physical Review Letters* **86**, 2909 (2001); S. Jespersen, I. Sokolov, and A. Blumen, *J. Chem. Phys.* **113**, 7652 (2000); H. Jeong, B. Tombor, R. Albert et al., *Nature* **407**, 651 (2000); M. Barthelemy and L. Amaral, *Physical-Review-Letters* **82** (15), 3180 (1999).

[3] Y. Chen, G. Rangarajan, and M. Ding, *Physical Review* **E 67**, 026209 (2003); A.S. Dmitriev, M. Shirokov, and S.O. Starkov, *IEEE Transactions on Circuits and systems I. Fundamental Theory and Applications* **44** (10), 918 (1997); P. M. Gade, *Physical Review* **E 54**, 64 (1996); J.F. Heagy, T.L. Carroll, and L.M. Pecora, *Physical Review* **E 50** (3), 1874 (1994); Gang Hu, Junzhong Yang, and Winji Liu, *Physical Review* **E 58** (4), 4440 (1998); L. Kocarev and U. Parlitz, *Physical Review Letters* **77**, 2206 (1996); Louis M. Pecora and Thomas L. Carroll, *International Journal of Bifurcations and Chaos* **10** (2), 273 (2000); C. W. Wu, presented at the 1998 *IEEE International Symposium of Circuits and Systems*, Monterey, CA, 1998 (unpublished); Chai Wah Wu and Leon O. Chua, *International Journal of Bifurcations and Chaos* **4** (4), 979 (1994); C.W. Wu and L.O. Chua, *IEEE Transactions on Circuits and Systems* **42** (8), 430 (1995).

[4] L.M. Pecora and T.L. Carroll, *Physical Review Letters* **80** (10), 2109 (1998).

[5] K. Fink, G. Johnson, D. Mar et al., *Physical Review* **E 61** (5), 5080 (2000).

[6] Unlike our earlier work we use a minus sign for the coupling terms so the definition of matrices from graph theory agree with the graph theory literature.

[7] .A. Turing, *Philosophical Transactions* **B 237**, 37 (1952).

[8] D.M. Cvetkovi´c, M. Doob, and H. Sachs, *Spectra of Graphs: Theory and Applications*. (Johann Ambrosius Barth Verlag, Heidelberg, 1995).

[9] B. Mohar, in *Graph Symmetry: Algebraic Methods and Applications, NATO ASI Ser. C*, edited by G.Hahn and G.Sabidussi (Kluwer, 1997), Vol. 497, pp. 225.

[10] Réka Albert and Albert-László Barabási, *Physical Review Letters* **84**, 5660 (2000).

[11] Réka Albert and Albert-László Barabási, *Reviews of Modern Physics* **74**, 47 (2002).

[12] Duncan J. Watts and Steven H. Strogatz, *Nature* **393** (4 June), 440 (1998).

[13] M. Barahona and L. Pecora, *Physical Review Letters* **89**, 054101 (2002).

[14] P. Erdös and A. Renyi, *Publicationes Mathematicae (Debrecen)* **6**, 290 (1959); P. Erdös and A. Renyi, *Bull. Inst. Internat. Statist.* **38**, 343 (1961).

In: New Research on Chaos and Complexity ISBN 1-60021-275-1
Editors: F. Orsucci and N. Sala, pp. 97-122

Chapter 6

CTML: A MARK UP LANGUAGE FOR HOLOGRAPHIC REPRESENTATION OF DOCUMENT BASED KNOWLEDGE

Graziella Tonfoni
DSLO University of Bologna
Bologna 40126, Italy

Author of the CPP-TRS methodology and of the CTML

ABSTRACT

This chapter is specifically meant to illustrate an articulated framework for information visualization based upon a set of conceptual tools meant to promote qualitative reasoning in documentation management.

The complexity of our information world today poses the immediate and urgent need for powerful conceptual tools, enabling individuals to cope with daily communicative activities which have become more and more complex.

On the other side decision makers in our interconnected world need to reach the top level of accuracy and be able minimize the risk of fuzziness and misinterpretation which may be caused by lack of originating context, where a certain information was first produced or by continuously shifting contexts throughout a diversity of media and cultures.

CTML, the Context Transport Mark up Language, (Tonfoni,1998i, 1999) is a high power derivative language of CPP-TRS, the Communicative

Positioning Program-Text Representation Systems (Tonfoni,1996i,1996ii), which is a widely tested and comprehensive methodology based upon a consistent set of dynamic visuals, to be used in combinations meant to teach individuals how to represent those cognitively different textual operations they perform, while they are writing and reading in multimedia environments.

CTML, which may definitely be viewed as a really advanced and a highly specialized sublanguage on its own; is also a mark up system meant empower already existing advanced information systems.

As a controlled language on its own it is meant to enable verification of asysmmetries which are likely to occur in interpretation of those very documents, which are most subject to continuous change, revision and further updates.

Keywords: Context identification, context transport, context reproduction, context verification

1.1. INTRODUCTION

Document management systems are crucial today. To ensure most effective performance, they need to incorporate highly specialized conceptual tools for identifying misunderstandings and for fixing interpretive mistakes. Such problems may be caused by significant changes occurring in our highly interconnected information environment, leading to unwished and unintended communication casualties and interpretation aberration phenomena.

Interpretation aberration phenomena are likely to occur for different reasons, such as information overload and loss of context and also as a consequence of uncontrolled perceptual changes, which are unintentionally triggered by too rapid data expansion and amplification produced by new information technologies.

A document may be viewed as white light, which is made out of a variety of colors.

An advanced document management system needs to provide users with special lenses that may allow various colors to all come up as the result of different kinds of refractions.

Documents and single paragraphs are most likely to absorb external energies, along with external and undesired information matter, which is part of the communicative environment, where such documents are first analyzed and then classified.

Our perceptual system is undergoing efforts of major proportion and major stress is generated as a consequence.

Information flowing and floating in huge and uncontrolled quantities is dramatically affecting our usual modes of perception. Meaning vibrations, caused by fuzzy communication waves and their order of magnitude and speed, may therefore alter dramatically an originating meaning even when just literally transported from one given context into another one.

Besides, random turbulences are generated within an information environment due to culture-bound interpretation attribution progressive waves, which may deeply affect the originating and intended meaning of each given document.

Any advanced document management system needs to be complemented by a whole set of tools and resources, which may serve the purpose of allowing information flows to come in; monitor, control, transfer information; filter, distinguish, discriminate, label information; verify, store, classify information; retrieve, declassify, package information; repackage, update, upgrade information; and finally select, reconfigure, deliver information.

Information detection may be envisioned as visualization enhancement made possible throughout a whole visual apparatus, enabling it to trace back communication patterns consistently and correctly.

Advanced document management systems will highly benefit from specifically targeted and fully enhanced information compilation capacities, making discrimination between verified and unverified documents actually possible.

A whole set of techniques for analyzing information meant to allow information capturing and reproduction of the closest approximation possible to the originating context have been derived from Tonfoni's theoretical framework and are incorporated within the CTML system, which is here synthetically illustrated.

Within such framework, information compilations of various kinds are made available and transportable as a consequence of a consensually shared set of criteria. Document management systems designed according to the highly articulated conceptual model here provided also provide a highly consistent visual screen and display for qualitative reasoning upon information throughout a highly specialised vocabulary to be shared interculturally.

More precisely: information compilations, visually enhanced, will provide: interpretive clues for understanding, parallely and synchronously supporting evidence for each statement made; interpretive cues showing that originating sources were in fact accessed and accurately referenced or referred to; interpretive labels based upon a set of preselected categories, which have been established as a consequence of an in-depth search and as the result of highly consistent knowledge accumulation, filtering and selection; interpretive labels based upon a set of paragraphs, which derive from an already summarized description of each item, and where paragraphs are complemented by comments and carry evaluation criteria along; interpretive cues indicating full continuity and consistency for further insertion of more summarized information, which is of relevance in order to better understand facts and events illustrated; interpretive cues in the form of a premise, meant to declare that more information is needed in order to know enough about a certain fact, event or phenomenon; and finally interpretive labels indicating a set of examples complemented by explanation and definition, showing to be a subset derived from a larger body of knowledge.

Representation of time actually needed for accurate perception, which will indicate clearly that time of reading will radically differ from time of writing, will also be displayed along with the document representation in progressively evolving stages.

1.1.1. On Document Based Knowledge

Continuous improvement within an organizational structure needs to be seen as strictly linked and dependent upon individuals' cognitive ability and teams' cooperative and communicative competence in monitoring continuously incoming information flows, some of which are synchronous and consistent, where others are asynchronous, contradictory up to fully fuzzy. Different information flows need to be carefully analyzed as to be made compatible.

Creating a conceptual platform for documentation sharing means being able to value each individual contribution to the production of knowledge within any organization, we know well as for most extended research in this area that a continuously shifting context of communication may result in a threatening and discouraging factor if not properly presented and then handled.

On the other hand, continuous sharing of information, coming in documental formats today, may be viewed positively and result into a unique opportunity for individuals to become active protagonists of a most significant process, such as the creation of a broader communication context.

Information managers today need to be aware all the time that they are working for the present and the future, and that only what has been consistently designed may be built in steady ways and is therefore likely to be stored as to remain available for further use. Documentation, which is conceived as to be turned into a collective memory based repository of organizational knowledge, is most likely to be productively accessed and reused at different times and for different purposes.

Enhanced data warehousing may therefore be consistently viewed as some kind of high performance knowledge management, where knowledge is the derivative product and the end result of a set of highly articulated processes of information selection and organization, made visible throughout a whole series of precisely defined transitional states, incorporating the history and evolution of each document. Within the framework, here illustrated, individuals' contributions will be stored and will therefore remain visible in their originating context, as such they will be accessible and available all the time.

Naturally occurring instability and sudden shifts of context and priorities, due to continuous change, are compensated for, by a flexible interpretive model for information visualization, which allows changes and context shifts to be reported, traced back and triggered together with the originating context through the use of a consensually agreed upon and shared code.

To summarize: each document contributor's visibility made possible throughout an enhanced text encoding system, will significantly gain out of consistent visualization of those very conditions in which the document was first generated and then revised, upgraded, updated up to even radically transformed.

1.1.2. Change: From Threat to Challenge

Even if interpreted as challenging and rewarding, there is no doubt that change and transitioning conditions within any organizational structure are perceived as threatening and need to be appropriately monitored.

Motivation and commitment by individuals or teams to just one stage of a process, which is presented and therefore perceived to be most likely either discontinued, or passed to different individuals or teams, is hard to keep.

Evidently individuals involved with a certain process perceive and resent instability either consciously or unconsciously, even if encouraged to feel committed to what they have been asked to do temporarily, to complete those tasks they have been assigned.

Results achieved by individuals may actually not even become part of the organizational memory and may not be used at all.

CTML allows a whole set of contributions, which are likely to remain invisible if waves of new information, often contradictory and fuzzy are allowed to flow in indiscriminately, to still remain fully available and visible as to facilitate tracing back of the originating sources and clear-cut decision making as for their reliability.

Individuals are today continuously encouraged to think of change as some real opportunity to learn for themselves first and as a true challenge to cooperatively interact and work in teams.

But it is also a fact that insecurity and instability are perceived as diminishing and frightening factors playing both against personal initiative and collaborative attitudes.

As a consequence of instability, information management subject to continuous change is characterized by the following features:

- increasingly wide amount of information coming in fuzzy, both synchronously and asynchronously;
- increasingly wide amount of contradictory and inconsistent data to be checked and verified continuously;
- increasingly frequent change of conditions of satisfaction and continuously occurring communicative context shifts.

Decision making based upon instable conditions is perceived as a fuzzy process linked to concepts such as arbitrariness and chance.

If decision makers may neither identify nor visualize relevant information as to be able to consequently envision a consistent model for taking action (Tonfoni,1998ii), will they not be able to make reasonable projections and predictions and will they be inclined, as a consequence, to resist full involvement and full commitment.

Decision making on most crucial and most delicate matters, when subject to change due to continuing modification of roles, played by single individuals and teams, is likely to become disturbed.

Information managers through the use of a CTML enhanced document management system are practically enabled to foresee, decide and disclose how long and to what extent they will actually be in charge of and effectively responsible for a certain process they have initiated.

They are this way more likely to resist providing their own experience to the new comers and they are definitely more keen to provide those knowledge based interpretive clues they have themselves developed in the course of the years to their collaborators or successors as to facilitate problem solving.

New comers are infact likely to miss most relevant knowledge and to be missing the context also as they are likely to be invited to take action unexpectedly and under fully different conditions and within a significantly modified context.

Such resistance is not to be referred exclusively to personality issues which are obviously there too, as it also derives from awareness that each change of context and protagonists does create a new situation, where a certain problem solving strategy, which may have been previously suggested and considered to be valid, may simply just not work any longer.

Documentation management is strictly bound to accuracy, motivation and responsibility shown by individuals involved in the process of creation and further development throughout a whole set of transitional states.

Lack of those elements radically affects the way information flows are perceived, monitored and channeled within an organizational structure. Collective memories show different structures and require complex representation systems. Information systems designers do have to tailor complexity of the representation according to organizational scale (Bowker and Star, 2000).

There needs to be a good match between the types of information collected in the form of documents and their repository and its basic mission and scope.

The framework here proposed is aimed toward creating and keeping optimized conditions for positively coping with continuous change, and for supporting effective decision making under unstable conditions by reinforcing the decision makers' roles and responsibilities and by providing both conceptual and practical tools meant to establish most favorable conditions for taking action.

2.1. Documentation Visualization as a Teaching and Learning Opportunity

Documentation (from Latin *documentum* : to be taught) is the result of whole set of textual operations performed by an individual or more individuals, working collaboratively in groups within an organizational structure.

Daily events, currently occurring interactions, occasional conversations as well as planned meetings establish the conditions for information to flow through different media in many and various ways and also create the context for an accurate understanding of the very culture of each organization and institution.

Information may therefore be viewed as coming in various flows and waves and also be filtered, categorized and organized, as to then become accessible and reusable for different purposes at different times.

As soon as information is found to be of relevance and becomes stabilized it is then turned into a document: at this very stage the need to store information in ways which may be made fully transparent becomes a major issue.

Information visualization based on both topic continuity and context consistency is a most fundamental process upon which accurate and timely decision making resides.

Availability and accessibility of contextual information coming in documental format is well supported by an enhanced labeling system, such as CTML is, which may help speed up effective search and retrieval in significant ways.

Each document or piece of a document- where by piece of document a consistent topic and context completed section of a document is indicated, which is recognized as a whole entity by itself, ready to be linked up to other documents, according to topic continuity and context consistency - may be labeled according to qualitative reasoning criteria upon the nature of information and effectively stored and retrieved.

Not only is it important to suggest statistical methods in order to be able to identify topical words in the form of keywords, but it is equally relevant to add interpretive clues as to be able to add qualitative reasoning clues on top.

By showing explicitly which kind of information each document contains, along with the originating context in which the information was first conceived, will the document become CTML screened and the full history of each paragraph may become visible.

By showing which kind of progressive revisions have in fact produced various documental transitional states and various outputs, and by indicating explicitly those upgrading and updating operations, which have been performed all along, it will also become possible to trace back individuals' contributions in reformulating different context shifts, which have occurred at various times and annotated as such.

Reshuffling pieces of information coming in written forms and turning them into documents which are stabilized information packages meant to incorporate CTML clues for interpretation, according to continuously changing scenarios, entails a specific competence and cannot be subject to arbitrary choice.

Not only is a very specific "context sensitivity" required, but also a consensually shared framework for interpretation needs to be made available and referred to in order to label documents and pieces of documents consistently, based upon a common understanding and naming of the different textual operations performed.

This is why a derivative controlled language such as CTML comes handy.

Each document in the process of being visualized and labeled and made ready to be reused represents a real learning opportunity for those individuals discovering, defining and finally showing those very same operations, which have been performed at each given stage.

As a consequence of the fully explicit illustration of a whole variety of actions taken upon information, some of which may also look contradictory, may the same document, become a teaching opportunity for those individuals and groups which need to access the document in order to gain knowledge about its own history as to process it further, by updating or upgrading it either locally or globally, according to the new circumstances.

According to a CTML perspective, will each single document carry its own attached history, while undergoing changes of various kinds.

Its visual annotation apparatus, such as displayed, may therefore be viewed as an enhanced attachment, keeping trace of the various transitioning stages, that may if not considered properly, affect its originating communicative value more or less radically.

Transitional states in a document are therefore to be considered those temporarily defined and stabilized states of information which provide evidence and support for a certain set of decision making processes which have occurred or are going to occur next.

Change applies to the conditions of satisfaction any communicative occurrence entails as well as to priorities and roles played by individuals.

Temporarily stabilized textual states will therefore provide context, evidence and visibility for each individual who is actively involved with the process.

Documentation organized and built up according to a CTML based perspective, may constitute a tremendously rich repository of collective memories, for each organization or institution which may want to invest into creating the optimized conditions for keeping the consistent interpretation of actions and activities available for understanding in the future as well.

Obviously, a consensually shared encoding system such as CTML is infact, needs to be made compatible with other information technologies, which are currently used by the organizational structure as to be really made transportable all throughout the different communicative situations occurring in the specific culture.

This is precisely the reason why a common visual system for encoding documentation has been found to be most flexible and handy to serve the purpose of transparency, clarity and crosscultural transportability.

2.1.1. Multiple Viewing and Complex Textual Visualization through Holographic Representation

Before becoming a reliable stabilized document, each text may be described, viewed and considered as a hologram.

Holograms are in fact a very good model to capture the various levels and layers of complexity each communicative interaction may entail, which then also reflects upon each paragraph.

Holograms are infact three dimensional representation systems which may go far beyond geographical mapping. In some ways, they are communication representation devices, which are likely to enhance human interpretive processes, just like the wheel is meant to enhance human motion capabilities not by imitation.

A documental hologram is a three dimensional representation meant to be perceived and processed according to a highly context sensitive set of techniques. Only a high level of accuracy and precision in the processing of each paragraph, which has been identified as particularly sensitive, will ensure the possibility of visualizing multiple and complex interpretation processes according to the document producer's point of view and perspective.

Two kinds of documental holograms are presented and described. They are precisely: transmission documental holograms and reflection documental holograms, which consistently apply to the descriptions of wider corpora, based upon joint interpretive efforts made by partners as to produce a stabilized body of document based evidence at the end.

Three dimensional visualization of different communicative positioning stages within the same document would in fact explode the number of interpretation processes, if larger interpretative patterns were not first identified and then indicated as such, in order to guarantee consistent document perception to occur at each given stage and time of observation on the document perceiver's side.

Differentiated stages of meaning may create a progressively shifting and dynamically changing document visualization and three dimensional representation of complex interpretation just on one page and may well serve both for exploiting meaning potential and for eliminating undesired ambiguity, before that very page becomes a stable document on which decision making may be based.

2.1.2. Visualization of Document Based Reasoning

It is by now very much accepted in the literature that a mental model for representing a variety of phenomena that involve knowledge representation and information processing (Johnson-Laird, 1983, Chi and Glaser and Farr, 1988) may become a useful device for information seeking (Marchionini, 1995).

More specifically diagrams are abstract graphic portrayals of the subject matter they represent (Lowe, 1993).

Correspondence between the diagram observer's and perceptor's reactions and the diagram author's intentions represents an interesting topic for exploration and further research as in the analysis of different types of individuals dealing with different kinds of problem solving tasks (Hegarty and Just and Morrison,1998).

A mental model for enlightening interpretation complexity will have to entail dynamic viewing of each text, where by text a basic linguistic entity is considered together with its own consistent communicative context, which may then, according to a multiple viewing and complex information visualization, be converted into a textual holographic representation.

A same model may guarantee for both complex and multiple contexts explosion of interpretation, which is generated by multiple context attribution and for accurate definition of one specific context, whenever multiple contexts possible attribution may result in a very undesirable condition, because it may generate confusion.

According to previous research work extensively carried on (Tonfoni,1996iii) , communicative context may be viewed as the combination of communicative function, communicative intention and communicative turn taking controlling each single paragraph or page.

Complexity evidently appears as an intrinsic aspect of any kind of documentation interpretation and management, at different levels and scales of course, which need to be taken into account.

Complexity may show at different ranks, like lexical complexity, which has a wide literature on its own.

What is really on focus here, is rather that kind of complexity which is directly associated with the different types of communicative contexts, which may control each page at different stages; they are in fact precisely related to communicative function, intention and turn taking.

Since communicative context controlling each paragraph, may vary significantly as for space and time, therefore creating different documentation perception possibilities and effects, it may become extremely useful to visualize potential interpretive values and sometimes even conflicting interpretation stages which may be analyzed in their combinations as well, also proceeding toward most different directions, even if generated and derived out of the same originating body of documentation.

Previously carried on analysis in terms of communicative pattern recognition, shows clearly how some links may be more predictable than others and how guessing and qualitative reasoning about documental further evolutions may be productively pursued.

According to such perspective, visualization comes in as an extremely powerful device, both as for interpretation and monitoring and for reducing distortion and consequently eliminating undesired ambiguity.

Visualizing complexity of interpretation first requires very sharp conceptual tools and it certainly demands that a previous selection has been made, meant to first identify different layers and levels of possible meaning, which should be first isolated and then categorized, as to be consistently handled.

A documental hologram becomes therefore not only an accurate metaphor for envisioning information exchanges, and for representing interpretation complexity, progressively shifting according to space and time perception, but also a very practical device for representing and processing a document according to a complex and multiple viewing based procedure, still in highly reliable and fully precise ways.

The interactive nature of each collaborative effort poses the need to identify two different kinds of documental holograms, "transmission documental holograms" and "reflection documental holograms" respectively.

Transmission documental holograms are monitored and controlled by the originating document producer, whereas reflection documental holograms are processed by the document receiver and are the result of the observer's interpretation, monitoring and perception.

Communicative context, meaning communicative function, intention and turn taking, may in fact be missing or be very ambiguous, even in the course of a collaborative documentation passing and exchange, given the gap both in time and space and the most likely variety of media and channels involved in our highly complex and highly interconnected world of today.

Documents are naturally subject to loss of context or context shift, which may result in possible ambiguity or multiple interpretation.

Documental holograms processing is therefore meant to make possible visualization of the many and complex ways in which communicative context would jeopardize interpretation and understanding if missing or left invisible or distorted.

3.1. On "Transmission Documental Holograms" and "Reflection Documental Holograms"

Transmission documental holograms should be organized by the originating document producers, who may want to make sure that the document is received, meaning perceived, charged by that very precise kind of communicative function, intention and turn taking, which they want themselves to assign to the document itself. If the documental hologram producer's main concern is to make sure that a stabilized and clearly defined context is conveyed together with text, the text will be turned into a three dimensional representation of each of its paragraph, charged with its specifically determined context.

Reading of the document may therefore be conceived in terms of progressive shifting by adopting a different perception, perspective and point of view at each given stage.

Transmission documental holograms may also serve the opposite purpose, which means that the document producer may decide to explode the context potential and multiple interpretation possibilities stage by stage by attributing to each paragraph a whole variety of plausible contexts.

If such process is to be undertaken, then a whole interpretation explosion will come out of it, allowing document perceivers to react in multiple ways as opposed to directing them toward a desired and preplanned interpretive direction, as in the previous case.

Reflection documental holograms derive from the need document perceivers have to define a certain context for themselves as for interpretation of each document received to be then sent back to the original producer with added comments.

By adding a context, are perceivers indicating what they think is relevant to the point.

This is meant to allow the original sender to add a new context or to modify the indicated one, therefore trying to optimize communication as much as possible by reducing the bandwidth and gap existing between producers and perceivers.

If ambiguity or high complexity of the original document is perceived, the reflection documental hologram may well represent such complexity, in order to have the same document be sent back to the original producer for selection and identification of most appropriate context for accurate interpretation.

A reflection documental hologram, in the form of a reply, may in this last case just represent the originating message together with one or many possible contexts for interpretation to be checked and confirmed or deleted by the original producer to be properly replied to by the perceiver according to the most suitable context attribution.

The whole concept of hologram is based upon effective recognition of intrinsic complexity of each written text, especially as space and time variables may play a significant role in document production and perception and temporary vacuum of context may occur up to context redundancy.

CTML based holographic techniques play an extremely relevant role in defining or selecting the most accurate context for consistent interpretation.

Keeping multiple contexts open may in some cases result in added value to the text, as we were taught to think about literary and poetic texts or under some other circumstances: it may actually constitute a quite significant problem for accurate interpretation of a very extended set of other kinds of texts.

Multiple possible contexts attribution may in fact result in noise and redundancy, therefore slowing down or even paralyzing a whole chain of consistent reasoning patterns based upon documental knowledge.

A snowball like effect on documentation interpretation and processing may actually increase the combinatorial possibilities of misinterpretation and misjudgment, and therefore dramatically alter the intended meaning of paragraphs, therefore jeopardizing consistent interpretation of the overall document.

In summary: a documental hologram is a high visibility device, and its illustrative power is high too. High visibility should not in any way be misunderstood and thought of as restrictive and constraining, rather as clarifying and disambiguating.

If multiple contexts are in fact to be kept, they may be represented and equally shown in a multiple holographic documental representation as intended by either producer or perceiver or both of them.

3.1.1. Enhanced Encoding Procedures for Documentation Visualization

The process of visualizing a certain document is here achieved throughout the pervasive and massive application of a consistent interpretive system, meant to describe and define different kinds of communicative actions taken by individuals and groups, to be finally stored electronically. Documental operations are here categorized and analytically defined in the process of progressive organization, each document is likely to undergo in the course of conversations and meetings to be finally turned into a consistent attachment to the document stored.

Once a consistently interpreted and appropriately packaged document or piece of document is labeled and recognized in its own originating context, it may then be reconfigured many times and more or less radically transformed. Such process may be activated after relevant clues have been extracted and visually represented by specific icons, supported here of course by the Context Transport Mark up Language, CTML, which are of four kinds, and are precisely the following ones:

- documental signs: meant to indicate the communicative function or type of a document or piece of a document;
- documental symbols: meant to indicate the communicative style of a document or piece of a document;
- documental turn taking symbols: meant to indicate the role and interplay between the document producer and the document reader;
- documental amplifier symbols: meant to coordinate wide sets of documents, which are characterized by topical continuity and context consistency.

By topical continuity, we mean to indicate documents focusing on the same topic, which may be either literally extracted as linear sequences out of the document or abstracted as a result of accurate interpretation and further adaptation at a more conceptual level.

By contextual consistency, we mean to indicate documents showing the same communicative context, which may be explicitly declared as to be easily retrieved.

Documental signs, which represent the various communicative functions, a document may convey paragraph by paragraph, are the following ones:

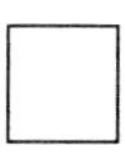

Square: for an informative document or piece of a document, which carries information about a specific event or fact, to be linked up with another document or set of documents made available, in order to extend topical continuity and context consistency.

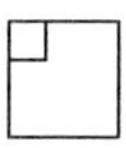

Square within the Square: for a summary of a certain document, which has been produced to reinforce contextual consistency between an original document and its own abstract.

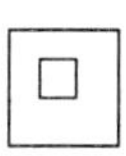

Frame: for a document or piece of document which is found to be analogous in content to other documents and previously stored cases meant to reinforce contextual consistency between and among different documents.

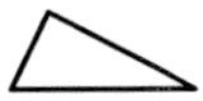
Triangle: for a memory and history generated out of a certain document meant to establish topical continuity with background information which has not been previously introduced because not available in a documental format.

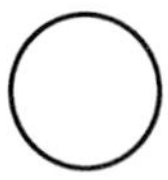
Circle: for a main concept conveyed by a certain document which has been abstracted and linked to other documents showing topical continuity. It is meant to reinforce topical keywords identification and to effectively link together documents which show the same keyword.

Grouped Semicircles: for main concepts which are abstracted out of an originating document and meant to establish both topical continuity and context consistency between the originating document and a set of topical keywords.

Semicircle: for a locally identified concept abstracted out of a piece of document and meant to reinforce context consistency by establishing further links to other documents which show the same keyword.

Inscribed Arcs: for indicating the need for an upgrade and update of a certain document; it indicates that a revision process is likely to occur, though it does not declare if such revision will be a major or a minor one.

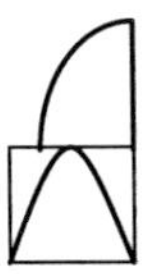
Opened Text Space: for indicating that an upgrade and update has indeed occurred within a certain document; it indicates that the document has now reached a new revision state as a consequence, though it does not declare if the revision has been a major or a minor one.

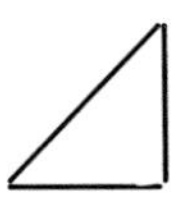
Right Triangle: for a comment made to a certain document or piece of a document coming in a non documental format where more contextual information is needed, which has to be derived from other external sources not previously available, based on topical continuity.

Documental symbols are meant to indicate communicative intentions and styles more locally within a certain document, sentence by sentence.

They are particularly useful in showing contributions made by individuals in the process of creation of a certain document and may be easily incorporated within the final document providing further interpretive clues which may significantly add to clarity and visibility.

Documental symbols, which represent different modes of information packaging activated at different times or at the same time, may be combined and used dynamically for repackaging purposes, because they effectively indicate documental transitional states by declaring explicitly the nature of those changes which have occurred or are likely to occur next.

Documental symbols are the following ones :

Describe: from Latin *describo*: write around. It means complementing the original document or piece of a document with as much information as maybe found interesting to add without any specific constraints.

It is represented by a spiral which starts from a central point –the middle point of the spiral indicating the original document- and proceeds toward expanding the document at various degrees, linking it with other documents or pieces of documents or information coming in from different sources and found to be relevant to facilitate the originating document interpretation.

In this specific information packaging mode, there is actually no need at all to follow any chronological order ; the spiral may be smaller or larger, depending on how much information the document producer - an individual or a team – may find relevant to add.

Define: from Latin *definio*: put limits. It means complementing the document with limited information about a very defined topic which has been previously selected and identified as the most relevant one, which is represented by the middle point of the square.

It indicates that there is a specific need to incorporate specific information about a relevant document or piece of a document which is made available.

Define means actually describe under specific constraints and implies accurate and most selective focusing on a very limited package of highly specific information.

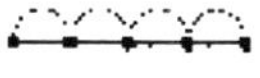

Narrate: from Latin *narro*: tell the story. It means complementing the document with various facts and events, which have been referred to in the originating context by following a logical and chronological order.

It indicates a set of major points or facts representing different diachronic stages which are strictly linked up together in a sequence.

The longer the story is the more narrative points are actually added according to the document producer's decision making.

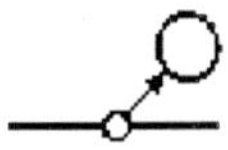

Point out: take a point out of a story chain. It means isolating a specific event or fact among those reported within a single document, focusing on just that one and adding more detailed information by expanding it significantly and linking it with other documents which have been found to be of relevance to that point.

It represents the specific point chosen and does obviously entail the need to look for more extended information provided and made available.

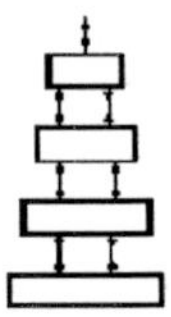

Explain: from Latin *explano*: unwrap, open up. It means that facts and reasons are given to support interpretation of a certain event within a certain document or piece of a document.

The document producer may start by indicating the originating cause and proceed toward showing the effects or start with effects and go back to the cause, according to what is found to be more significant.

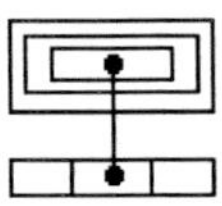

Regress: from Latin *regredior*: go back. It means that more information about a certain topic presented within the document is absolutely needed as to gain a deeper understanding.

It represents a specific topic focusing process and an in depth information expansion, which is activated only for that precise topic.

The document reader may want to consider if further information is needed and ask for availability of additional resources.

Inform: from Latin *informo*: put into shape, shape up. It means that any document is the result of some information organization and that the very specific document indicated is organized in the most unconstrained way, therefore subject to many and various kinds of reformulations.

It usually leads toward two different kinds of further specification, which are respectively conveyed by the "inform synthetically" and the "inform analytically" indication:

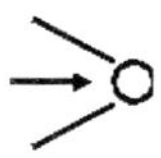

"inform synthetically" means departing from a larger document or set of documents and proceed toward a summary related to a specific topic, identified as being the most relevant one emerging from the originating document.

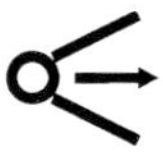

"inform analytically" means departing from a given document or limited set of documents to expand toward further documents or add more information, which needs to be previously converted into a documental format not yet available.

Reformulate: from Latin *reformo/reformulo*: change shape and shape again.

It means changing the kind of information packaging which was adopted before and substituting a certain information request with a different one still related to the same document.

It may turn into a more or less radical transformation of the originating document according to a precisely defined request or set of requests.

Express: from Latin *exprimo*: push out and press out.It means adding personal opinions and individual feelings related to facts and events within a certain document; it indicates the most subjective mode of information organization, which is openly recognized to be bound to very personal evaluations, judgments and emotional states.

Documental turn taking symbols are meant to define the mode of accessing and reading the document, requested at each given time; they are suggested by the document producer to be followed by the document user; they are the following ones:

Major Scale: it shows that literal interpretation is needed and that those pieces of documents indicated and marked off, should be extracted and quoted literally the way they were first organized.

Minor Scale: it shows that accurate interpretation may need a further process of abstraction and that pieces of documents indicated and marked off may undergo significant reconfiguration processes up to high level conceptualization.

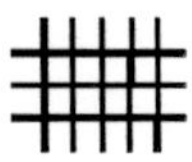

Open or Unsaturated Rhythm: it shows that accessing the document may lead the user toward incomplete interpretation of those facts and events, which are presented. It is meant to suggest the user to access more documents and various kinds of sources which are made available.

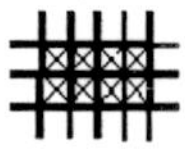

Tight or Saturated Rhythm: it shows that accessing the document will lead the user toward complete interpretation of those facts and events which are presented. It is meant to suggest the user to stick to the interpretation provided, though access to other sources is still available to support evidence.

Documentation amplifier symbols are complementary and may be added after the previously illustrated ones have been used; they apply to sets of documents and larger documentation territories and indicate specific operations, which are to be performed to connect sets of documents, which have been previously encoded and accurately stored.

They are the following ones:

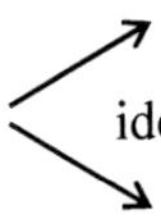

Choose: it is meant to represent the dynamic process of first identifying and then deciding between optional contexts for interpretation, which are mutually exclusive, given a certain set of documents.

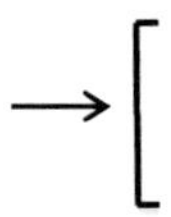

Identify:it is meant to represent definition of a more specific context within a broader context for interpretation of a set of documents; it naturally occurs before "search" and "select".

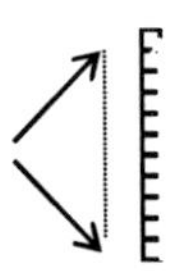

Search: it is meant to represent the dynamic process of choosing among different contexts for interpretation of a set of documents which are many and compatible as to find the most appropriate one.

Select: it is meant to represent multiple contexts which may evolve either synchronously or asynchronously and may be modified once a certain decision making process has been performed and be stored and kept as an example.

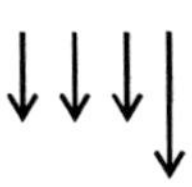

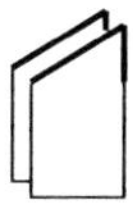 **Copy/Replicate**: it is meant to represent the dynamic process of duplication and repetition of a certain context, which, if lost, would radically jeopardize understanding and accurate interpretation of a set of events and facts described and explained by a set of documents.

→ **Ahead**: it is meant to represent the progression of a certain set of documents which are linked together by context consistency or harmoniously shifting contexts.

← **Back**: it is meant to represent the need to go back to delete and replace the originating context which has radically shifted in the course of various transition states, such that, if not eliminated, would indeed affect consistent interpretation of a whole set of documents.

→ ← **Conflict**: it is meant to represent an emerging inconsistency and incompatibility between various context attributions to a set of documents which needs to be cleared as to proceed toward any further interpretation.

The documental notational system here illustrated in its various components may be applied at different layers and at various levels of complexity and is meant to underline the fundamental role and responsibility of the encoding individual and encoding team.

It is obvious that interpretive clues assigned to wide information territories will result into the production of a large amount of diversified documentation which will constitute some kind of enhanced and threedimensional geographic mapping system, carrying along its own legend, which has a direct impact upon the culture of the organizational structure it is applied to.

3.1.2 On Continuity and Change

Interpretation consistency, in spite of continuously occurring shifts of context and turn taking by individuals may still be possible. By retrieving the different layers of interpretation provided to a certain fact or set of facts, also reflecting directly upon further stages of information packaging, resulting then into a final documental format, evidence for statements made may be supported in the very form of an attachment for readers, also enhanced by a whole set of indications about the specific team members who have been put in charge and are therefore responsible for interpretation and mapping.

Just like any geographic map will only show those features which are relevant according to the nature and purpose assigned to the map itself, the same way of thinking may be extended to the domain of documentation mapping which need to be designed according to different kinds of priorities which may come out progressively as a consequence of the continuously shifting contexts.

Energy and time dedicated to quite an expensive process, such as enhanced encoding of each state of information is when packaged into a paragraph, is anyway fully cost effective because it is meant to provide an enormous amount of examples and in-house knowledge, that will remain extensively available for future generations to come.

Any further process of verification may tremendously gain and profit out of a dynamic repository which is meant to expand all the time still keeping track of the history of each given document.

Sense of responsibility as well as pride are directly connected and linked to motivation and sense of identification.

If it is a fact that not being able to foresee a future reward for a project may be a source of deep frustration for each individual involved, it is equally true that valuing each process in all of its stages, by keeping records about each individual's involvement and contribution, is a fundamental step for resourceful rethinking within a continuously changing information affluent society.

A document, which could not be completed in the past, as a certain project may have been disconnected, may turn out to be profitably triggered later on, providing therefore an example, which may serve the purpose to inform newcomers of previous work done by individuals involved.

According to such perspective, individuals initiative and authorship recognition as for documentation production is harmoniously incorporated as part of the enhanced encoding process and may reflect directly upon history of each document attached, which is meant to explicitly declare and illustrate roles played by single individuals or by highly recognizable teams.

Document management systems when CTML enhanced also represent a powerful carrier, transporting through time and space accurately filed and verified documentation viewed as the output result of individuals and teams, whose contributions are made visible according to what they decide should infact be made visible and explicit, as carrying along the context and cultural flavour in which the information packaging and repackaging did actually occur.

CTML enhanced repositories are meant to encourage active protagonism, to support consistent turn taking and to facilitate accurate decision making at each given time, ensuring individuals that each contribution provided and found to be of relevance will not be lost.

The concept of CTML enhancement in the documentation production lifecycle, resides upon the concept of harmonious integration of multiple information resources.

Rethinking of the document production lifecycle along those lines may result in a natural process very much custom friendly as memories of single contributors may be made available for further and multiple access and on line verification.

Reinventing documentation formats according to the CTML document design conceptual infrastructure is also meant to ensure continuity with the past contributions as scattered pieces of information may be made compatible and transportable.

Providing a consensually shared framework for tagging and labelling documents also represents a very constructive solution, as to cope with useless dispersion up to dissipation of relevant knowledge, which may have consolidated in the course of the years, as a consequence of the individual and collective effort of thinking and rethinking of day by day events.

4. Conclusions

Reinventing documentation formats does not mean reinventing the wheel again and again, pretending it was not there before. A CTML enhanced encoding system is based upon accurate analysis, rethinking and retracing "how the wheel was first invented and then progressively modified and refunctionalized according to continuously changing priorities and diversified needs as to best suit different kinds of evolving vehicles."

To extend the metaphor a bit further, document design innovation within any organization or institution needs to be bound to a deep understanding of the already existing tradition and of the very reasons, why such tradition was first created and evolved in various ways. New vehicles may only be conceived and designed as a consequence of observation, analysis and most accurate interpretation of the prototypes and further models which were reconfigured afterwards.

CTML, also a document annotation system, has been designed specifically as to enhance content and context visibility and intended to provide a means for information and knowledge management.

Mission of this long lasting research effort, extensive testing and verification, has been and is to be able to greatly enhance the accuracy of the process of information retrieval and of knowledge conveyance and cognition. CTML, also as a controlled language, has a whole variety of applications to information and knowledge retention, information and knowledge classification, contextual refinement of stored information, and contiguous context linking of web-based information.

Qualitative reasoning about the nature of information displayed within a document is made possible and may be further incorporated within the automation process of reviewing and redacting documents. Energy and time to be dedicated to the process of document redesigning according to the new conceptual formatting procedures have shown evidently to be fully compensated by the increased level of accuracy reached in the interpretation process.

REFERENCES

Bowker, G.C.; Star, S.L. (2000), *Sorting things out: classification and its consequences*, Cambridge, Mass., M.I.T. Press

Chi, M.T.H.; Glaser, R., Farr, H.J. (1988), *The nature of expertise*, Hillsdale, N.J., Lawrence Erlbaum

Hegarty, M.; Just, M.A., Morrison, I.R. (1988), Mental models of mechanical systems: individual differences in qualitative and quantitative reasoning, in "*Cognitive Psychology*", 20,191-236

Johnson-Laird, P.N. (1983), *Mental models: toward a cognitive science of language, inference and consciousness*, Cambridge, Cambridge University Press

Lowe, R.K., (1993), Diagrammatic information: techniques for exploring its mental representation and processing, in "*Information Design Journal*", 7., 1, 3-17

Marchionini, G. (1995), *Information seeking in electronic environments*, Cambridge, Cambridge University Press

Tonfoni, G. (1996i), *Communication Patterns and Textual Forms*, Exeter, U.K., Intellect

Tonfoni, G. (1996ii), A visual news processing environment, in Artificial Intelligence Review, *Special Issue on Integration of Natural Language and Vision Processing: Recent Advances*, edited by P. McKevitt, Kluwer Academic Publishers, Amsterdam, The Netherlands, 10,3-4,33-54

Tonfoni, G. (1996iii), On visual text planning, processing and programming, in "*Machine Learning and Perception*", edited by G. Tascini, F. Esposito, V.

Roberto and P. Zingaretti, Series in Machine Perception and Artificial Intelligence, vol.23, World Scientific, Singapore,23. 99-106

Tonfoni, G. (1998i), Intelligent control and monitoring of strategic documentation: a complex system for knowledge miniaturization and text iconization, in "*Proceedings of the ISIC/CIRA/ISAS 98 Conference*", National Institute of Standards and Technology, U.S. Department of Commerce, Gaithersburg, Maryland, U.S, 869-874.

Tonfoni, G. (1998ii), *Information Design: The Knowledge Architect's Toolkit*, Scarecrow Press, Lanham, Maryland, U.S.

Tonfoni G.(1999), On augmenting documentation reliability through communicative context transport, in *The Proceedings of the 1999 Symposium on Document Image Understanding Technology*, SDIUT99, Annapolis MD, 283-286

ACKNOWLEDGMENTS

Anytime I present my visual system, in one of its many possible applications, I cannot forget to mention those people, who have supported me, by creating the conditions for my work to become actually visible and to circulate as to be practically used.

First of all, I want to thank Marvin Minsky, Toshiba Professor at the Massachusetts Institute of Technology who has always encouraged me to go ahead with my research and has, in the course of many years, created for me the opportunities for presenting the output results of my work and for discussing it with him and colleagues at the Massachusetts Institute of Technology, especially when I was a Research Scholar at the Artificial Intelligence Laboratory.

I am very grateful to Masoud Yazdani, Professor at the University of West England Bristol and President of Intellect for having been most supportive in having my books timely published and who has shown significant interest and a clear understanding of my work in the course of the years.

I want to thank colleagues at the School of Engineering and Applied Science at The George Washington University for having provided opportunities for my work to be fully illustrated, and particularly Chairman and Dean Lile Murphree.

I also want to thank Dean Tom Mazzuchi, and Professors Michael Stankoski and Dianne Martin also directors of the Cyberspace Policy Institute, and Professors Rachel Heller, Robert Lindeman, Richard Scotti and Julie Ryan. Christopher Heikiman and Richard Stoler have provided me with valuable professional support as for my activities, and workshops in the Washington area.

I then want to acknowledge very valuable encouragement in pursuing this work, given to me by Dr. Pete Daniel, Curator in the Division of the History of Technology at the National Museum of American History at the Smithsonian Institution in Washington D.C., who has also carefully read this paper and keeps records of my work done in the Washington area.

In: New Research on Chaos and Complexity ISBN 1-60021-275-1
Editors: F. Orsucci and N. Sala, pp. 123-135

Chapter 7

SUSTAINABILITY AND BIFURCATIONS OF POSITIVE ATTRACTORS

Renato Casagrandi[*,†] ***and Sergio Rinaldi***[†,‡]
[†] Dipartimento di Elettronica e Informazione, Politecnico di Milano, Milano, Italy
[‡] Adaptive Dynamics Network, International Institute for Applied Systems Analysis, 2361 Laxenburg, Austria

ABSTRACT

In this paper we show how sustainability can be rigorously defined by making reference to the positivity of the attractors of a dynamical system. Consistently, the sustainability analysis with respect to various system and policy parameters can be performed by using specialized software for the study of the bifurcations of nonlinear dynamical systems. By means of an example concerning the tourism industry, we show how the analysis can be systematically organized and how easy it is to interpret the results of the numerical bifurcation analysis.

1. INTRODUCTION

The notion of sustainability is, nowadays, one of the most pervasive (if not invasive) in all political debates. The idea of sustainability emerged in the late sixties and can be traced in some pioneering scientific work, like that of Hardin

---

[*] Corresponding author: casagran@elet. polimi.it

(1968) or those that inspired the Club of Rome (Forrester, 1971), and of various sociopolitical movements (http://greenpeace.org, http://wwf.organd http://zero growth.org are few among the hundreds).

A great number of studies followed the pioneering stage and gave rise to *conferences*, as the 1992 UN Commission on Environment and Development conference in Rio de Janeiro, *journals* (like International Journal of Sustainable Development, Sustainable Development and World Ecology, Environmental Modeling and Assessment, Environment and Development Economics, Ecological Economics and others), *books* (Clark and Munn, 1987; Costanza, 1991; Wackernagel and Rees, 1995; Dodds, 2000; Starke, 2002), *laws*, as the European Directives 337/85 and 11/97 for the Environmental Impact Assessment or the 42/2001 for the Strategic Environmental Assessment, and *international agreements*, as the still unattended Kyoto protocol (see http://unfccc. int/cop3/for details). The main result of this huge effort is that people and governments are now much more sensitive than thirty years ago to the problem of long term survival of the world. However, despite this success, the issue of sustainability is still missing a simple and clear theoretical framework. This is very unfortunate, because in the absence of unified theories and methods of analysis any issue, no matter how important it is, becomes vague and anoising and, in the long run, discourages young scientists from investing their skillness.

Since sustainability refers to the possibility of keeping alive forever all meaningful social and natural compartments of an evolving system (from towns to continents) it is clear that any *formal* definition of sustainability must refer to the long term behavior of some appropriate dynamical system. Thus, one should *a priori* expect that the analysis of sustainability with respect to the parameters characterizing the system (e.g. latitude, resource availability, population, . . .) or its government (e.g. standards on emissions, environmental taxation schemes, subsidies, . . .) can be performed through the study of the bifurcations of the attractors of a mathematical model mimicking the evolution of the real system. This is, indeed, the thesis of this article, which has the ambitious target of establishing a bridge between an important issue (sustainability) and a basic chapter of modern mathematics (bifurcation analysis of dynamic systems).

The paper is organized as follows. In the next section two components of sustainability, called *profitability* and *compatibility*, are defined with reference to an abstract model of the system. The first component takes into account only the social compartment of the system, while the second is only concerned with the environmental aspects. From these definitions it follows that the study of profitability and compatibility can be carried out through the bifurcation analysis of the attractors of the model. However, not all the attractors of the system are involved, but only those which are "positive" with respect to the social or to the environmental variables. Then, in the third section, sustainability is defined by putting social and environmental aspects at the same level of importance. This definition is in line with the theory of conflict resolution in multiobjective analysis and it is not as partisan as others proposed by many economists and environmentalists. Again, from our definition it follows that a bridge can be established between sustainability and bifurcation theory. Finally, an entire section is devoted to highlight through an example the meaning of the various definitions given in the paper and to show how the bridge established with bifurcation theory can allow one to systematically and effectively discuss sustainability once a model of the system is available.

2. Profitability and Compatibility

We now assume that the time evolution of the variables involved in the problem under consideration is described by a set of ordinary differential equations (ODE). We also assume that the state vector can be partitioned in three subvectors x, y and z of dimensions n_x, n_y and n_z, respectively, and that the variables x_i, i=1, ..., n_x and y_j, j=1, ..., n_y are indicators of social and environmental value. For example, the variables x_i could be measures of employment, welfare, health or education in a given nation, while the variables y_j could be abundances of some plant or animal species in a forest, air quality in various towns, water quality in some rivers and lakes and so on.

All these variables x_i and y_j are typically non-negative, because they represent, directly or not, densities or biomasses. The equations describing the evolution of x_i and y_j over time are simply *conservation equations* involving the balance between inputs and outputs. Moreover, the input and output rates in the balance equations are, with almost no exception, expressed in terms of net per capita rates. In other words, the rate of variation of x_i ($dx_i /dt = \dot{x}_i$) is the product of the abundance x_i and the net growth rate per capita f_i, which is a function of all variables. All this brings to the conclusion that the model can be assumed to have the following general form

$$\dot{x}_i = x_i f_i(x, y, z, p, q) \qquad i = 1, \ldots, n_x \tag{1}$$

$$\dot{y}_j = y_j g_j(x, y, z, p, q) \qquad j = 1, \ldots, n_y \tag{2}$$

$$z_k = H_k(x, y, z, p, q) \qquad k = 1, ..., n_z \tag{3}$$

where f_i and g_j are net growth rates per capita, z_k are variables of no direct social and environmental interest, and p and q are constant parameters which identify the characteristics of the system (altitude, structure of transportation networks, fleet dimension, . . .) and of its management (emission standards, fishing quotas, subsidies for tourism development, . . .).

The particular form of eqs. (1,2) (sometimes called Kolmogorov's form) is such that the non-negativity of the variables x_i and y_j is preserved forever if it is guaranteed at the initial time t=0, i.e. the space $x_i \geq 0$, $y_j \geq 0$ for all i, j is an invariant set. For physical reasons, in the following we will always refer to this invariant set even if we do not say it explicitly. Given the pair (p, q), i.e. given system (1-3), all its attractors in the above invariant set are uniquely identified (even if not always easily computable). These attractors can be many. Some of them can be not positive – i.e. x_i =0 and y_j =0 for some (i, j)–, while others are positive with respect to x(i.e. the attractor is characterized by $x_i > 0 \,\forall\, i$) and/or with respect to y (i.e., $y_j > 0 \,\forall\, i$). From now on, an attractor which is positive with respect to x will be called x-positive, and similarly for y, while an attractor which is positive with respect to x and y will be called (x, y)-positive. The existence of an x-positive attractor guarantees the possibility that all compartments of social interest remain alive forever. From an economic viewpoint, this means that the system has the possibility of producing profits forever. This justifies the following definition.

Definition 1 *A pair (p,q)is profitable if system (1-3) has at least one x-positive attractor.*

From this definition it follows that the points (p*, q*) of the boundary of the profitability region in the space (p, q) are bifurcation points of system (1-3). In fact, given a boundary point (p*, q*) there exist pairs (p, q) infinitely close to (p*, q*) for which system (1-3) has an attractor characterized by $x_i > 0 \,\forall\, i$. By varying the parameters p and q, the attractor must cease to exist or to be x-positive at (p*, q*). In the first case the attractor has a catastrophic bifurcation at (p*,q*); for example, if the attractor is an equilibrium it disappears through a saddle-node bifurcation or through a subcritical Hopf bifurcation, if it is a limit cycle it disappears through a tangent bifurcation of cycles or through a homo-clinic bifurcation, and so on for more complex attractors (i.e., tori and strange attractors). In contrast, if the attractor does not disappear at (p*, q*) but loses its positivity with respect to x at that point, then system (1-3) has a transcritical bifurcation at (p^*, q^*). This is a consequence of the Kolmogorov's form of eq. (1) which has the constant solution x_i =0 for all values of p and q Notice that transcritical bifurcations are generic in Kolmogorov's systems (Kuznetsov, 1995).

It is worth noticing that not all bifurcations of system (1-3) are involved in performing the profitability analysis of a system. In fact, bifurcations of attractors which are not positive with respect to x have nothing to do with profitability. The same is true for x-positive attractors which remain such while undergoing a non-catastrophic bifurcation (e.g. a supercritical Hopf bifurcation). Finally, if the system has multiple x-positive attractors, a catastrophic or transcritical bifurcation of one of them does not imply the loss of profitability, which is guaranteed by the remaining x-positive attractors.

It is important to notice that the persistence of all social characteristics is not always guaranteed in a profitable system. In fact, in such a system, besides an x-positive attractor $A'(p, q)$, there can also be another attractor $A''(p, q)$which is not positive with respect to x, namely an attractor characterized by x_i =0 for some i. In such a case, a sufficiently strong perturbation, like a political scandal, an epidemics or a war, can move in a very short time the state of the system from the attractor A' into the basin of attraction of the alternative attractor A''. Thus, after the perturbation has ceased, the system will tend toward A'' and lose some of its social characteristics. It is therefore lecit to distinguish between safe and risky profitability as follows.

Definition 2 *A profitable pair (p, q) is safe if all the attractors of system (1-3) are x-positive, and risky otherwise.*

Analogous considerations hold for the environmental characteristics of the system: when it is possible to preserve them forever, we say that the system is compatible.

Definition 3 *A pair (p, q) is compatible if system (1-3) has at least one y-positive attractor.*

We can also distinguish between safe and risky compatibility as follows.

Definition 4 *A compatible pair (p ,q) is safe if all the attractors of system (1-3) are y-positive, and risky otherwise.*

Of course, the boundary of the compatibility region in the space (p, q) enjoys the same properties pointed out for the boundary of the profitability region.

Thus, points (p^*, q^*) belonging to that boundary are catastrophic or transcritical bifurcation points of system (1-3).

3. SUSTAINABILITY

In order to give a definition of sustainability which is not as partisan as the two previous ones, we pretend that both the socio-economic and the environmental compartments of the system persist forever. In other words, our formal definition of sustainability is the following.

Definition 5 *A pair* (p, q) *is sustainable if system (1-3) has at least one* (x, y) *positive attractor.*

Notice that a sustainable system is both profitable and compatible, while the converse is not always true. In fact, a system could have many attractors, some x-positive and some y-positive, without having one (x, y)-positive attractor.

As far as risk is concerned, in the case of sustainability one can distinguish between economic and environmental risk as follows.

Definition 6 *A sustainable pair* (p, q) *is safe if all attractors of system (1-3) are* (x, y)*-positive, and risky otherwise. Moreover, a risky sustainable pair* (p, q) *is at economic [environmental] risk if one of the attractors of system (1-3) has* $x_i = 0$ *for some I* [$y_j = 0$ *for some j*].

Of course, the boundary of the sustainability region in the space (p, q) can be determined through bifurcation analysis by looking, in particular, at the catastrophic and transcritical bifurcations of the (x, y)-positive attractors. Moreover, also the boundaries separating safe and risky systems are composed of bifurcation points.

4. AN EXAMPLE

The example we present in this section concerns tourism sustainability. The model is an extension of a simpler model described in detail in Casagrandi and Rinaldi (2002). The problem is rather abstract and refers to a hypothetical site characterized by four variables: number of tourists (x), environmental quality (y), and capital, subdivided into two quotas (z_1 and z_2) measuring the value (amount) of structures producing touristic services of different nature (e.g., cultural and recreational). The model has the following form

$$\dot{x} = x\left[\mu_y \frac{y}{y+\varphi_y} + \mu_z \frac{\lambda z_1 + (1-\lambda)z_2}{\lambda z_1 + (1-\lambda)z_2 + \varphi_z x + \varphi_z} - \alpha x - a\right] \quad (4)$$

$$\dot{y} = y\left[r\left(1+\frac{y}{K}\right) - \beta_1 z_1 - \beta_2 z_2 - \gamma x\right] \quad (5)$$

$$\dot{z}_1 = -\delta z_1 + \nu\varepsilon x \quad (6)$$

$$\dot{z}_2 = -\delta z_2 + (1-\nu)\varepsilon x \quad (7)$$

At the right-hand-side of eq. (4), the first term is the flow of tourists attracted by the environmental quality of the site, while the second is the flow of tourists attracted by services (with λ and (1- λ) representing the relative preferences for the two classes of services); the third term is negative and specifies how quickly the tourists abandon the site when it is crowded, while the last term is the basic rate at which tourists abandon the site. If the site is absolutely not attractive ($y = z_1 = z_2 = 0$) and there is no crowd, the tourists decay as *exp*(*-at*). Of course, the rate a of this exponential decay is higher if there are many other interesting touristic sites. It is therefore reasonable to assume, as done in the following, that the parameter a is a measure of the competition of the alternative touristic sites.

In eq. (5) the first term says that in the absence of touristic activities ($x = z_1 = z_2 = 0$)the environmental quality recovers to a carrying capacity K in accordance with a logistic equation, while the other terms represent the environmental impacts due to the supply of touristic services and to the presence of tourists.

Equations (6) and (7), which are linear, say very simply that the structures needed for producing services would become obsolete at an exponential rate δ if part of the profit (proportional to the number of visiting tourists) would not be reinvested in the service sector.

Assume that we are interested in the sustainability of the tourism industry and we want to discover the impact of the parameters ε (reinvestment of the touristic agents, for simplicity called *investment*) and a (competition of alternative touristic sites, called *competition*). Moreover, suppose that we also like to detect the effect on sustainability of λ and ν, which represent the preference of the tourists and of the agents for the first class of services. For this we must perform a bifurcation analysis of model (4-7) with respect to ε, a, λ, ν. The first step can be the study of the case $\lambda = \nu = 0.5$, in which the two classes of services are not distinguishable. Figure 1 shows the bifurcation diagram in the space (ε, a)for the reference parameter values given in the caption. The diagram has been obtained using specialized software for bifurcation analysis based on continuation techniques (Khibnik et al., 1993; Doedel et al., 1997).

The attractors are either equilibria or limit cycles and there are only five types of bifurcations (Kuznetsov, 1995), that is

TC_{eq} = transcritical bifurcation of equilibria;
SN_{eq} = saddle-node bifurcation of equilibria;
$PLASN_{eq}$ = saddle-node bifurcation of equilibria in the plane y=0, $z_1 = z_2$;
HOPF = supercritical Hopf bifurcation;
HOM = homoclinic bifurcation.

Figure 1 shows that there are two codimension-2 bifurcation points: a Bogdanov-Takens (see point *BT*, where a saddle-node, a Hopf, and a homoclinic bifurcation curve merge) and a degenerate saddle-node (see, point *TCSN*, where a transcritical and a saddle-node bifurcation curve merge). The space (ε, a)is partitioned into 10 regions, each one characterized by a specific set of attractors. From Definition 1 it follows that the system is not profitable only in region 8, i.e. the tourists can be permanently present on the site if the competition of the alternative sites is not too high. Similarly, from Definition 3 it follows that the system is not compatible only in region 10, i.e. the environment is necessarily jeopardized by tourism activities if agents reinvest a lot (ε high) and alternative sites are not very competitive (a low). Finally, from Definition 5 it follows that the system is not sustainable in regions 8, 9, and 10, i.e. not only where it is not profitable (region 8) and where it is not compatible (region 10), but also in region 9 where the system is both profitable and compatible.

Figure 1 explicitly shows that not all bifurcations are boundaries of the profitability, compatibility, and sustainability regions. For example, the Hopf bifurcation curve is not a boundary of these regions.

The same bifurcation curves can be used to further partition the profitability, compatibility, and sustainability regions into safe and risky subregions. For example, Figure 2, which has been extracted from Figure 1, shows the subregions in which sustainability is safe and those in which it is at economic and/or environmental risk. From Figure 2 one can immediately conclude that the system can be sustainable and safe only if the alternative sites are not too competitive and the agents are not too aggressive in reinvesting their profits. An increase of competition first gives rise to an economic risk and then to the collapse of the tourism activities. Vice versa, an increase of the aggressivness ε of the agents first generates some environmental risk and finally jeopardizes the environment.

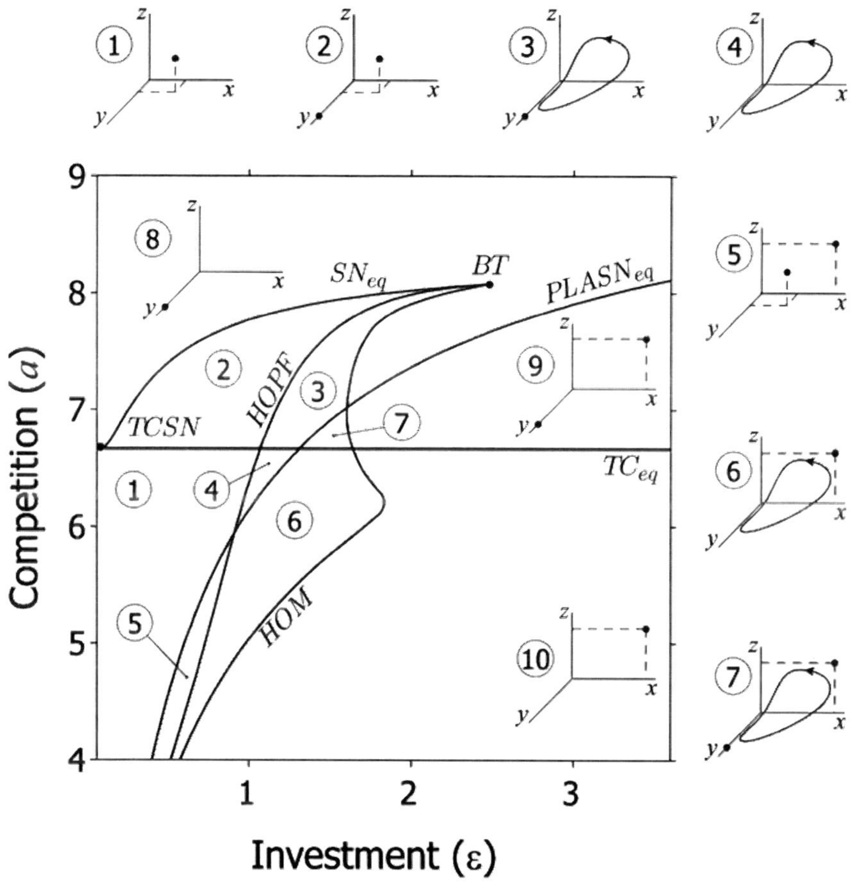

Figure 1: Bifurcation diagram of model (4-7) in the parameter space (ε, a) for the case $\lambda = v = 0.5$. The attractors in each region of the parameter space are sketched in the three dimensional space (x, y, $z = z_1 + z_2$). Other parameter values are as follows: $r = K = \alpha = \gamma = \varphi_z = 1$, ${}_y = \mu_z = 10$, $\varphi_y = \beta = 0.5$, $\delta = 0.1$

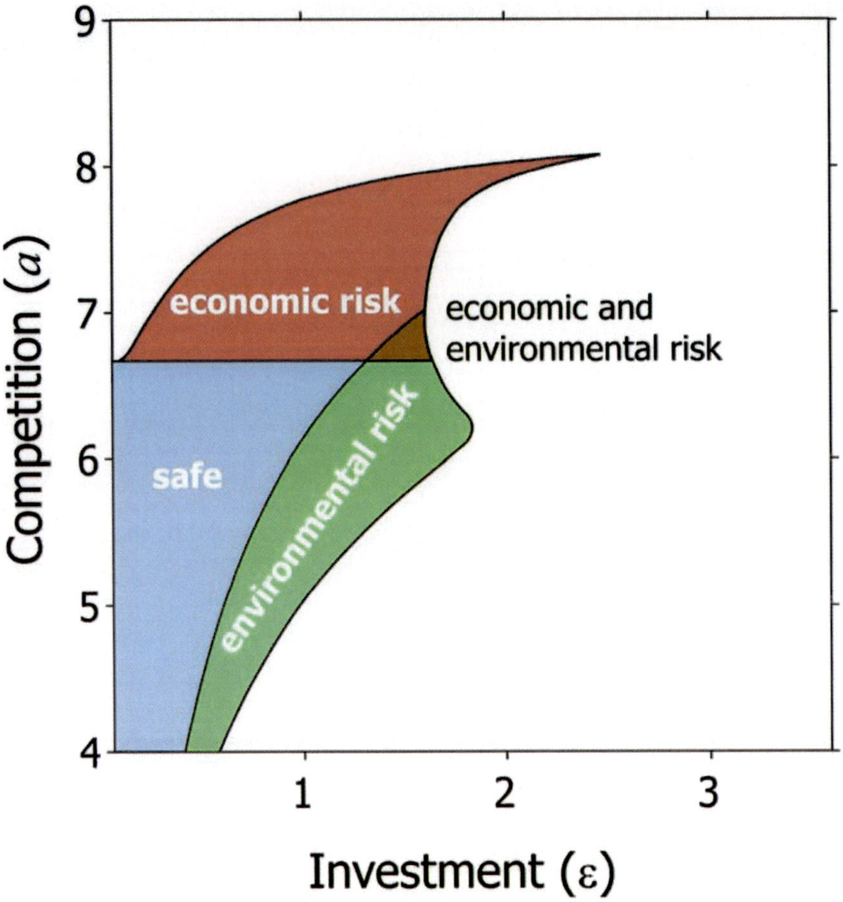

Figure 2: Sustainability diagram of model (4-7) with respect to investment and competition. Parameter values as in Figure 1.

Once the analysis for the symmetric case $\lambda = \nu = 0.5$ has been performed, the parameters λ and ν can be relaxed and the same software can be used to complete the analysis through continuation, starting from Figs. 1 and 2. Two bifurcation diagrams produced in this way are shown in Fig. 3. The first refers to the case in which tourists are more interested in the first kind of services ($\lambda = 0.8$), while agents invest primarily in the second class of services ($\nu = 0.2$). The effect, with respect to the symmetric case, is an increase of the safe sustainability region and of that at environmental risk. In the opposite case, i.e. when agents adapt to the preferences of the tourists ($\lambda = \nu = 0.2$), the above regions shrink and the system becomes more robust with respect to the competition of the alternative sites. All these properties, as well as others that could be easily obtained through continuation, are extremely useful for deriving qualitative but meaningful conclusions on tourism sustainability.

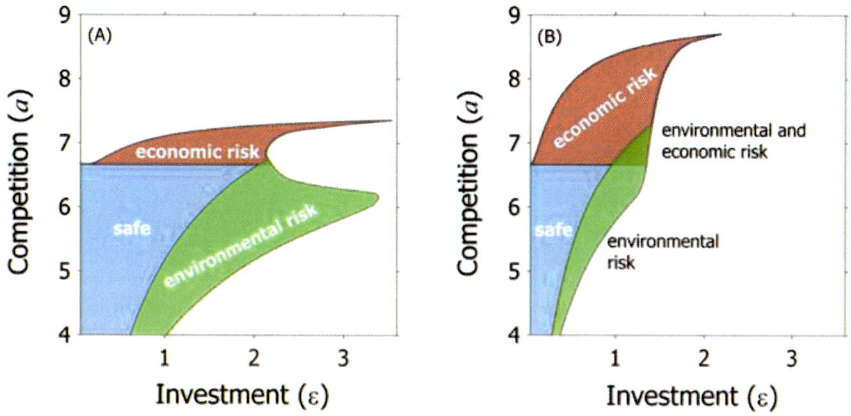

Figure 3: Sustainability diagram of model (4-7) with respect to investment and competition in the asymmetric cases where λ =0.8and ν =0.2(left panel) or λ =ν =0.2(right panel). Other parameter values as in Figure 1.

5 Conclusions

Defining sustainability is a very difficult task, since everyone has different perspectives on what should be sustained. In this article we have tried to look at two of the most commonly visited viewpoints. According to the majority of socio-economical scientists, humans well-being and prosperity are issues of primary importance and should be saved with the highest priority. For them, Nature is often the lower trophic level at the expenses of which we can happily survive and grow. Other researchers, especially conservation biologists and philosophers, claim that we are but a very marvellous species that must coexist with other natural beauties, like animals and plants, clear lakes, pristine seas or green mountains with pure air. Many human activities often sounds to them like disturbances that interfere with the organic life of the biosphere. Mediation between these viewpoints is impossible. However, everyone of us has some economic interests in her/his everyday life *and* has the profound necessity of interacting with an alive environment, not simply for exploiting it in the present or in the near future.

We decided to be extreme here. Compatible is every policy that does not completely destroy the environment. Profitable is every policy that ensures some persistent income, no matter how little it can be. Evidently no conservation biologist or economist would easily accept these crude definitions, but surely they will agree that a non-compatible or non-profitable policy cannot be *realistically* sustained in the long run. If the policy is such that a situation which is simultaneously compatible and profitable does exist, thus we can hope to sustain a community. Of course, the income can be unacceptably too little or the environmental conditions can be too contaminated. This is matter of specific considerations or personal judgement and it is hard to see how a formal method applicable in general to any dynamical model can solve the issue. The method we have proposed, however, helps in selecting the set of sustainable policies in a rigorous way. Trashing the unsustainable can be a good starting point, indeed, in extremely complex situations.

ACKNOWLEDGMENTS

The authors are grateful to Francesco Botti who has performed the numerical analysis for the proposed example.

REFERENCES

Casagrandi R and Rinaldi S (2002). A theoretical approach to tourism sustainability. *Conservation Ecology* 6(1):13. [online] URL: *http//www.consecol.org/vol6/iss1/art13.*

Clark WC and Munn RE, editors (1987). *Sustainable Development of the Biosphere*. Cambridge: Cambridge University Press.

Costanza R., editor (1991). *Ecological Economics: The Science and Management of Sustainability*. New York: Columbia University Press.

Dodds F, editor (2000). *Earth Summit 2002: A New Deal*. London: Earthscan.

Doedel EJ, Champneys AR, Fairgrieve TF, Kuznetsov YA, Sandstede B, and Wang XJ (1997). AUTO97: Continuation and bifurcation software for ordinary differential equations. Technical report, Department of Computer Science, Concordia University, Montreal, Canada. (Available by FTP from ftp.cs.concordia.ca in directory pub/doedel/auto).

Forrester JW (1971). *World Dynamics*. Cambridge, MA: MIT Press.

Hardin G (1968). The tragedy of the commons. *Science* 162:1243–1248.

Khibnik AI., Kuznetsov YA, Levitin VV, and Nikolaev, EV (1993). Continuation techniques and interactive software for bifurcation analysis of odes and iterated maps. *Physica D* 62:360–371.

Kuznetsov YA (1995). *Elements of Applied Bifurcation Theory* . New York: Springer Verlag.

Starke L, editor (2002). *State of the World 2002*. New York: Norton.

Wackernagel M and Rees W, editors (1995). *Our Ecological Footprint: Reducing Human Impact on the Earth* . NGabriola Island, Canada: New Society Publisher.

INDEX

A

B

D

E

F

G

H

I

Q

R

S

T